THE ART OF LOGIC
IN AN ILLOGICAL WORLD

Also by Eugenia Cheng

Beyond Infinity

How to Bake Pi

THE ART OF LOGIC
IN AN ILLOGICAL WORLD

EUGENIA CHENG

BASIC BOOKS
New York

Hachette Book Group supports the right to free expression and the value of copyright. The purpose of copyright is to encourage writers and artists to produce the creative works that enrich our culture.

The scanning, uploading, and distribution of this book without permission is a theft of the author's intellectual property. If you would like permission to use material from the book (other than for review purposes), please contact permissions@hbgusa.com. Thank you for your support of the author's rights.

Basic Books
Hachette Book Group
1290 Avenue of the Americas, New York, NY 10104
www.basicbooks.com

Printed in the United States of America

Originally published in hardcover and ebook by Profile Books in the United Kingdom in July 2018

First US Edition: September 2018

Published by Basic Books, an imprint of Perseus Books, LLC, a subsidiary of Hachette Book Group, Inc. The Basic Books name and logo is a trademark of the Hachette Book Group.

The Hachette Speakers Bureau provides a wide range of authors for speaking events. To find out more, go to www.hachettespeakersbureau .com or call (866) 376-6591.

The publisher is not responsible for websites (or their content) that are not owned by the publisher.

Library of Congress Control Number: 2018948797

ISBNs: 978-1-5416-7248-2 (hardcover); 978-1-5416-7250-5 (ebook)

LSC-C

10 9 8 7 6 5 4 3 2 1

For my parents

who taught me both logic and intuition.

CONTENTS

INTRODUCTION

WOULDN'T IT BE HELPFUL if everyone were able to think more clearly? To tell the difference between fact and fiction, truth and lies?

But what is truth? Is the difference between "truth" and "untruth" always that simple? In fact, is it *ever* that simple? If it is, why do people disagree with each other so much? And if it isn't, why do people ever agree with each other at all?

The world is awash with terrible arguments, conflict, divisiveness, fake news, victimhood, exploitation, prejudice, bigotry, blame, shouting, and miniscule attention spans. When cat memes attract more attention than murders, is logic dead? When a headline goes viral regardless of its veracity, has rationality become futile? Too often, people make simple and dramatic statements for effect, impact, acclaim, and to try and grab some limelight in a world where endless sources are competing relentlessly for our attention all the time.

But the excessive simplifications push us into fabricated black and white situations when everything is really in infinite shades of gray and indeed multi-colors. Hence we seem to live with a constant background noise of vitriol, disagreement, and tribes of people attacking other tribes, figuratively if not for real.

Is all hope lost? Are we doomed to take sides, be stuck in echo chambers, never agree again?

No.

There is a lifebelt available to anyone drowning in the illogic of the modern world, and that lifebelt is logic. But like any lifebelt, it will only help us if we use it *well*. This means not only understanding logic better, but also understanding emotions better and, most importantly, the interaction between them.

Only then can we use logic truly productively in the real human world.

The discipline of mathematics has carefully honed the techniques of logic, and as a research mathematician I come from this background. I believe we can learn from the techniques and insights of mathematics, because it's about constructing rigorously logical arguments and then convincing other people of them. Math isn't just about numbers and equations: it's a theory for justification. It provides a framework for having arguments and is so successful that in math people actually agree regularly upon conclusions.

There is a widespread myth that mathematics is all about numbers and equations, and that its usefulness in the world is in all the places we use numbers in life. The myth continues with the mistaken idea that the whole point of math is to turn life situations into equations, and solve them using math. While this is one aspect of math it is a very narrow and limiting view of what mathematics is and what it does. From this perspective we have "pure mathematics" as a rarefied field of esoteric symbols, far away from the real world, only able to interact with the real world via a chain of intermediaries:

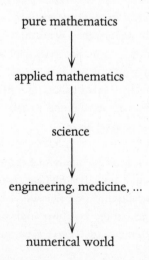

pure mathematics

applied mathematics

science

engineering, medicine, ...

numerical world

Instead we should branch out from this narrow, linear, incomplete view of math to use it in a much broader and hence more widely applicable sense. Mathematics in school may well be mostly about numbers and equations, but higher-level mathematics is about *how to think*, and in this way it is applicable to the entire human world, not just the part involving numbers.

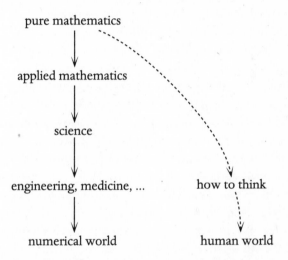

Mathematics helps us think more clearly, but it doesn't tell us *what* to think, and nor will I in this book. Contrary to how it might seem, math isn't about right and wrong, and nor are most arguments. They're about the *sense in which* something is right and wrong, depending on world views. If people disagree, it's often a result of different points of view stemming from different fundamental beliefs, not that one is right and the other is wrong.

If the idea of mathematics and logic seems remote and abstract to you, you are right: mathematics and logic *are* remote and abstract. But I will argue that the abstraction has a purpose, and that broad applicability is one of the powerful consequences. The remoteness of mathematics also has a purpose: taking a step back enables us to focus on important principles and think more clearly about them before putting the messy human details back in.

And we *will* put those details back in. We will analyse and illuminate messy, controversial, divisive issues such as sexism, racism, privilege, harassment, fake news and more. Logic does not resolve these issues, but clarifies the terms in which we should have the discussion. So I certainly won't be telling you what the conclusion of those arguments should be, but rather, how to have the argument in the first place.

In this book I will show the power of logic but also its limitations, so that we can use its power responsibly as well as effectively. In the first part I'll look at how we use logic to verify and establish the truth, by building clear, irrefutable arguments. In the second part I'll look at where logic breaks down and can't help us any more. As with any tool, we should not try to push logic beyond its limits, and so in the last part of the book I'll look at what we should do instead. Crucially, we need to bring emotions in too, first to find our way to the logic and then to convey it to others. Logic makes our arguments rigorous but emotions make them convincing. In the so-called "post-truth" world, truth seems to be accessed largely by emotions rather than logic. This sounds like it might be bad for rationality, but I will argue that it doesn't have to be a bad thing, as long as emotions are working *with* logic rather than working against it.

Emotions and logic do not have to be enemies. Logic works perfectly in the abstract mathematical world, but life is more complicated than that. Life involves humans, and humans have emotions. Here in this beautiful and messy world of ours we should use emotions to back up logic, and logic to understand emotions. I firmly believe that when we use emotions and logic together, each working to their own strengths and not beyond them, we can think more clearly, communicate more effect-ively, and achieve a deeper and more compassionate under-standing of our fellow human beings. That is the true art of logic.

PART I

THE POWER OF LOGIC

1

WHY LOGIC?

THE WORLD IS A VAST and complicated place. If we want to understand it we need to simplify it. There are two ways to make something simpler – we can forget parts of it, or we can *become cleverer* so that things that used to seem incomprehensible become clear to us. This book is about the role that logic can and should play in this process of understanding. It is about how logic can help us see and understand the world more clearly. It is about the light that logic shines.

Logic involves *both* aspects of making something simpler. Forgetting details is the process of abstraction, where we find the essence of a situation and concentrate on it for a while. Importantly, we must not forget critical details – that would be simplistic rather than illuminating. And we only do it temporarily, so that we are not claiming to have understood everything, but rather, a central core on which all further understanding can be rooted.

We will begin, in this chapter, by discussing why logic is a good foundation for all understanding, and what role logic can possibly have in a world of illogical human beings.

ACCESSING TRUTH

All areas of research and study are concerned with uncovering truths about the world. It might be the earth, the weather, the outer reaches of the universe, birds, electricity, brains, blood, people thousands of years ago, numbers, or something else. Depending on what you're studying, you'll need different ways of determining what is true, and of convincing other people that you're right. Anyone can make claims about what they think is true, but unless they back up their claims in some way, maybe nobody will believe them, and rightly so.

So different subjects have different ways of accessing truth.

Scientific truth is determined using the scientific method, which is a clearly defined framework for deciding how likely something is to be correct. It usually involves forming a theory, gathering evidence, and rigorously testing the theory against the evidence.

Mathematical truth is accessed by logic. We can still use emotions to feel it, understand it, and be convinced of it, but we use only logic to verify it. This distinction is important and subtle. In a way, we do *access* mathematical truth using emotions, but it doesn't count as true until we have *verified* it using logic.

The word "logic" is sometimes thrown around in disagreements to try and give an argument weight. "Logically this has to be true", or "Logically that can't be right", or "You're just not being logical!" The word "mathematically" also gets thrown around in this way. "Mathematically, they just can't win the election." Unfortunately these uses are often meaningless, more of a last-ditch way to try and shore up a weak argument. While the abuse of these words devalues them and so makes me sad, I am an optimist and so I choose to find something heartening in this as well: I am heartened to think that at some level people know that logic and mathematics are irrefutable and so should conclusively end an argument. If their names are being taken in vain to vanquish an opponent, at least this means there is some sense in which their power is being acknowledged.

Instead of simply lamenting the misunderstanding of logic and mathematics, I choose to address it, in the hope that their power might actually be used to good purpose. That is why I've written this book.

ADVANTAGES OF USING LOGIC

One of the main reasons to have a clear framework for accessing truth is to be able to agree about things. This seems very radical in a world in which people seem to revel in disagreeing with other people as much as possible. And it even happens in sport, when fans get angry with a decision that a referee has made, although the referee is supposed to be simply applying the agreed rules.

I remember watching the Oxford–Cambridge boat race one year when the boats clashed dangerously, and Cambridge was penalized. As a Cambridge person I was outraged because it looked to me that Oxford had obviously veered into them deliberately, so it looked like Oxford's fault. I thought the referee was in a conspiracy with Oxford and was being biased. However, instead of railing against this assumed conspiracy I looked up some expert commentary to try and understand what had happened. I learned that during the race up the river Thames, an imaginary line is drawn along the middle of the river, and each boat has priority on their side of the river. This means one boat can leave a lot of space, perhaps when taking a bend, and "lure" the other boat across the line. Then the boat with priority can deliberately veer into the boat that crossed the line, knowing they won't get the penalty. Is it morally right? Who's fault is that, really? We'll unravel questions of blame and responsibility in Chapter 5.

This idea of a clear framework for reaching consensus is also a bit like how medical diagnosis works. The medical profession tries to make a clear checklist so that a diagnosis is unambiguous, and so that diagnoses are made consistently by different people across the profession.

The idea of logic is to have clear rules so that conclusions can be drawn unambiguously and consistently by different people. This is wonderful in theory, and perhaps here "in theory" means in the abstract world of mathematics. Mathematics has a remarkable ability to make progress. Philosopher Michael Dummett writes in *The Philosophy of Mathematics*:

> Mathematics makes a steady advance, while philosophy continues to flounder in unending bafflement at the problems it confronted at the outset.

Why are mathematicians able to reach agreements about what is true? And why do those things remain true thousands of years later, where other subjects appear to be continually refining and updating their theories? I believe the answer lies in the robustness of logic. That is its great advantage.

There are also some disadvantages of the logical world. One is that you can't win arguments just by yelling loudly. Of course, this is only a disadvantage if you like winning arguments by yelling loudly, which I actually don't. But unfortunately plenty of people do, so they don't like the logical world. And they don't like the fact that in the logical world they can't get the better of a small, soft-spoken, uncool person like me. Because in the logical world strength doesn't come from big muscles, large amounts of money, or sporting prowess. It comes from sheer logical intellect.

Another disadvantage of the logical world is that you don't really have your feet on the ground any more, because we're no longer in the concrete world. It can sometimes feel like you're floating around in mid-nowhere, but I find that this is quite a pleasant sensation once you get used to it. The key, as with putting the first human in space, is to be able to come back again. In this book we are not just going to float around in the abstract world for fun. We are going to come back to earth and use powerful logical techniques to disentangle real, relevant, urgent arguments about the state of society. We are going to show that accessing the logical abstract world enables us to get further in the real world, just as flying through the sky enables us to travel further and faster in real life. In essence, this is the whole point of mathematics.

WHAT MATH IS AND ISN'T

There are many misconceptions about mathematics. These probably come from the way math is presented at school, as a set of rules you have to follow to get the right answer. The right answer in school math is usually a number. When proof finally comes into it it's often in the form of geometry, where "logical proofs" are constructed using peculiar facts to prove other pointless results such as that if you have some configuration of lines crossing each other in a bunch of different places then one angle over here turns out to be related to another one over there.

Warning/reassurance: this example is a spoof and can't be solved.

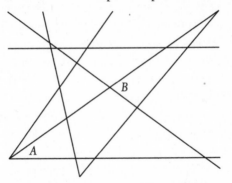

"Show that angle *A* is half of angle *B*."

You then have a series of tests and exams thrown at you, where you're asked to do a whole series of these pointless exercises under an arbitrary time pressure. If you make it through all that and somehow still believe you like math, you might go to university to do math, where the whole thing is likely to happen all over again except harder. If you make it through *that* and still think you like math, you might do a PhD and start doing research. Here, finally, mathematics starts to resemble what I think mathematics really is. Not a series of hoops to jump through, not an attempt to get the "right answer", but a world to explore, discover and understand: the logical world.

At this point some people realize that the thing they liked about "math" up until then was jumping through hoops and getting the right answer. They liked being able to get the right answer easily, and once they get to this exploratory world of math they run away.

Other people held on to a love of math throughout unfortunate school experiences because they somehow knew that math was going to be better and more exciting than that in the end, when they got to do research. Educator Daniel Finkel calls this being "inoculated" against school math classes. I was inoculated against them by my mother, who showed me that math was more than what we did at school. Some people are inoculated by an excellent math teacher – it sometimes only takes one teacher, for one class, to effect the inoculation and convince students that no matter what happened to them before and what happens to them after that class, math will be open-ended and fascinating if they pursue it long enough.

So what is this "real math" that we usually only meet when we start research? What is mathematics? Many people think it is "the study of numbers", but it is so much more than that. I gave a talk about symmetry to an elementary school in Chicago, and one little boy complained afterwards "Where are the numbers?" I explained that math is not just about numbers, and he wailed "But I want it to be about numbers!"

The rules of scientific discovery involve experiments, evidence and replicability. The rules of mathematical discovery do not involve any of those things: they involve logical proof. Mathematical truth is established by constructing logical arguments, and that is all.

My favorite way to think about math is: it is the study of how things work. But it's not the study of how any old things work: it's the study of how *logical* things work. And it's not any old study of how logical things work: it's the *logical* study of how *logical* things work.

Mathematics is the logical study of how logical things work.

Any research discipline has two aspects:

1. *what* it studies, and
2. *how* it studies it.

The two are linked, but in mathematics they are particularly cyclically linked. Usually the objects we're studying determine how we're going to study them, but in math the way we're studying them also determines what we can study. The method we're using is logic, and so we can study any objects that behave according to the rules of logic. But what are those? That is the subject of the first part of this book.

RULES

Different games and sports have different rules for deciding who is best in an unambiguous way. Personally, I am more comfortable with the ones that are very clear cut, like: who passed the finish line first, or who jumped over the highest bar without knocking it down. Other sports like gymnastics or diving seem more complicated, confusing and ambiguous if they require a panel of judges to make decisions according to certain criteria. The criteria are supposed to be set up to be unambiguous and to remove human judgement from the situation. But if they were truly unambiguous the judges would never disagree, and we wouldn't need a whole panel of them.

But even the apparently clear-cut sports have plenty of rules. If we look more closely at the 100 m sprint or the high jump, we see that there are rules about false starts, drug use, who is allowed to participate as a woman, who is allowed to participate as able-bodied, and more.

One problem with logic, as with sport, is that the rules can be baffling if you're not very used to them. I am pretty baffled by the rules of American football. Americans often assume that this is because I'm British and so used to "soccer" football, but in fact I'm baffled by that kind of football too. Although it does at least involve moving a ball around with your feet, so I understand that much.

We need to be clear what the rules of a sport are before we can really start playing it, and we need to be clear what the rules of logic are before we can really start using it. As with sport, the more advanced we get, the more deeply we'll have to understand the rules and all their subtleties. It's an effort, but the more we understand about the underlying principles of logic, the better and more productive arguments we'll be able to have.

A THEORY FOR ARGUING

The internet is a rich and endless source of flawed arguments. There has been an alarming gradual increase in non-experts dismissing expert consensus as elite conspiracy, as with climate science and vaccinations. Just because a lot of people agree about something doesn't mean there is a conspiracy. Many people agree that Roger Federer won Wimbledon in 2017. In fact, probably everyone who is aware of it agrees. This doesn't mean it's a conspiracy: it means there are very clear rules for how to win Wimbledon, and many, many people could all watch him do it and verify that he did in fact win, according to the rules.

The trouble with science and mathematics in this regard is that the rules are harder to understand, so it is more difficult for non-experts to verify that the rules have been followed. But this lack of understanding goes back to a much more basic level: different uses of the word "theory". In some uses, a "theory" is just a proposed explanation for something. In science, a "theory" is an explanation that is rigorously tested according to a clear framework, and deemed to be statistically highly likely to be correct. (More accurately, it is deemed statistically unlikely that the outcome would occur without the explanation being correct.)

In mathematics, though, a "theory" is a set of results that has been proved to be true according to logic. There is no probability involved, no evidence required, and no doubt. The doubt and questions come in when we ask how this theory models the world around us, but the results that are true inside this theory must logically be true, and mathematicians can all agree on it. If they doubt it, they have to find an error in the proof; it is not acceptable just to shout about it.

It is a noticeable feature of mathematics that mathematicians are surprisingly good at agreeing about what is and isn't true. We have open questions, where we don't know the answer yet, but mathematics from 2000 years ago is still considered true and indeed is still taught. This is different from science, which is continually being refined and updated. I'm not sure that much science from 2000 years ago is still taught, except in a history of science class. The basic reason is that the framework for showing that something is true in mathematics is logical proof, and the framework is clear enough for mathematicians to agree on it. It doesn't mean a conspiracy is afoot.

Mathematics is, of course, not life, and logical proofs don't quite work in real life. This is because real life has much more nuance and uncertainty than the mathematical world. The mathematical world has been set up specifically to eliminate that uncertainty, but we can't just ignore that aspect of real life. Or rather, it's there whether we ignore it or not.

Thus arguments to back something up in real life aren't as clean as mathematical proofs, and that is one obvious source of disagreements. However, logical arguments should have a lot in common with proofs, even if they're not quite as clear cut. Some of the disagreement around arguments in real life is unavoidable, as it stems from genuine uncertainty about the world. But some of the disagreement is avoidable, and we can avoid it by using logic. That is the part we are going to focus on.

Mathematical proofs are usually much longer and more complex than typical arguments in normal life. One of the problems with arguments in normal life is that they often happen rather quickly and there is no time to build up a complex argument. Even if there were time, attention spans have become notoriously short. If you don't get to the point in one momentous revelation, it is likely that many people won't follow.

By contrast a single proof in math might take ten pages to write out, and a year to construct. In fact, the one I'm working on while writing this book has been eleven years in the planning, and has surpassed 200 pages in my notes. As a mathematician I am very well practiced at planning long and complex proofs.

A 200-page argument is almost certainly too long for arguments in daily life (although it's probably not that unusual for legal rulings). However, 280 characters is rather too short. Solving problems in daily life is not simple, and we shouldn't expect to be able to do so in arguments of one or two sentences, or by straightforward use of intuition. I will argue that the ability to build up, communicate and follow complex logical arguments is an important skill of an intelligently rational human. Doing mathematical proofs is like when athletes train at very high altitude, so that when they come back to normal air pressure things feel much easier. But instead of training our bodies physically, we are training our minds logically, and that happens in the abstract world.

THE ABSTRACT WORLD

Most real objects do not behave according to logic. I don't. You don't. My computer certainly doesn't. If you give a child a cookie and another cookie, how many cookies will they have? Possibly none, as they will have eaten them.

This is why in mathematics we forget some details about the situation in order to get into a place where logic does work perfectly. So instead of thinking about one cookie and another cookie, we think about one plus one, forgetting the "cookie" aspect. The result of one plus one is then applicable to cookies, as long as we are careful about the ways in which cookies do and don't behave according to logic.

Logic is a process of constructing arguments by careful deduction. We can try to do this in normal life with varying results, because things in normal life are logical to different extents. I would argue that nothing in normal life is truly entirely logical. Later we will explore how things fail to be logical: because of emotions, or because there is too much data for us to process, or because too much data is missing, or because there is an element of randomness.

So in order to study anything logically we have to forget the pesky details that prevent things from behaving logically. In the case of the child and the cookies, if they are allowed to eat the cookies, then the situation will not behave entirely logically. So we impose the condition that they are not allowed to eat the cookies, in which case those objects might as well not be cookies, but anything inedible as long as it is separated into discrete chunks. These are just "things", with no distinguishable characteristics. This is what the number 1 is: it is the idea of a clearly distinguishable "thing".

This move has taken us from the real world of objects to the abstract world of ideas. What does this gain us?

ADVANTAGES OF THE ABSTRACT WORLD

The advantage of making the move into the abstract world is that we are now in a place where everything behaves logically. If I add one and one under exactly the same conditions in the abstract world repeatedly, I will always get 2. (I can change the conditions and get the answer as something else instead, but then I'll always get the same answer with those new conditions too.)

They say that insanity is doing the same thing over and over again and expecting something different to happen. I say that logic (or at least part of it) is doing the same thing over and over again and expecting the same thing to happen. Where my computer is concerned, it is this that *causes* me some insanity. I do the same thing every day and then periodically my computer refuses to connect to the wifi. My computer is not logical.

A powerful aspect of abstraction is that many different situations become the same when you forget some details. I could consider one apple and another apple, or one bear and another bear, or one opera singer and another opera singer, and all of those situations would become "1 + 1" in the abstract world. Once we discover that different things are somehow the same, we can study them at the same time, which is much more efficient. That is, we can study the parts they have in common, and then look at the ways in which they're different separately.

We get to find many relationships between different situations, possibly unexpectedly. For example, I have found a relationship between a Bach prelude for the piano and the way we might braid our hair. Finding relationships between different situations helps us understand them from different points of view, but it is also fundamentally a unifying act. We can emphasize differences, or we can emphasize similarities. I am drawn to finding similarities between things, both in mathematics and in life. Mathematics is a framework for finding similarities between different parts of science, and my research field, category theory, is a framework for finding similarities between different parts of math. We will show the efficacy of thinking in terms of relationships in Chapter 6.

When we look for similarities between things we often have to discard more and more layers of outer details, until we get to the deep structures that are holding things together. This is just like the fact that we humans don't look extremely alike on the surface, but if we strip ourselves all the way down to our skeletons we are all pretty much the same. Shedding outer layers, or boiling an argument down to its essence, can help us understand what we think and in particular can help us understand why we disagree with other people.

A particularly helpful feature of the abstract world is that everything exists as soon as you think of it. If you have an idea and you want to play with it, you can play with it immediately. You don't have to go and buy it (or beg your parents to buy it for you, or beg your grant-awarding agency to give you the money to buy it). I wish my dinner would exist as soon as I think of it. But my dinner isn't abstract, so it doesn't. More seriously, this means that we can do thought experiments with our ideas about the world, following the logical implications through to see what will happen, without having to do real and possibly impractical experiments to get those ideas.

HOW DO WE GET TO THE ABSTRACT WORLD?

Getting to the abstract, logical world is the first step towards thinking logically. Granted, in normal life we might not need to go there quite so explicitly in order to think logically about the world around us, but the process is still there when we are trying to find the logic in a situation.

A new system was recently introduced on the London Underground, where green markings were painted onto the platforms indicating where the doors would open. Passengers waiting for the train were instructed to stand outside the green areas, so that those disembarking the arriving train would have space to do so, instead of being faced with a wall of people trying to get on. The aim was to try and improve the flow of people and reduce the terrible congestion, especially during the rush hour.

This sounds like a good idea to me, but it was met with outcry from some regular commuters. Apparently some people were upset that these markings spoilt the "competitive edge" they had gained through years of commuting and studying train doors to learn where they would open. They were upset that random tourists who had never been to London before would now have just as much chance of boarding the train first.

This complaint was met with ridicule in return, but I thought it gave an interesting insight into one of the thorny aspects of affirmative action: if we give particular help to some previously disadvantaged people, then some of the people who don't get this help are likely to feel hard done by. They think it's unfair that only those other people get help. Like the absurdly outraged commuters, they might well feel miffed that they are losing their "competitive edge" that they feel they have earned, and they think that everyone else should have to earn it as well.

This is not an explicitly mathematical example but this way of making analogies is the essence of mathematical thinking, where we focus on important features of a situation to clarify it, and to make connections with other situations. In fact, mathematics as a whole can be thought of as the theory of analogies. We will use analogies to pivot between apparently unrelated situations throughout this book, and will provide a detailed analysis of the role of analogies in Chapter 13. Finding analogies involves stripping away some details that we deem irrelevant for present considerations, and finding the ideas that are at the very heart making it tick. This is a process of abstraction, and is how we get to the abstract world where we can more easily and effectively apply logic and examine the logic in a situation.

To perform this abstraction well, we need to separate out the things that are inherent from the things that are coincidental. Logical explanations come from the deep and unchanging meanings of things, rather than from sequences of events or personal decisions and tastes. The inherentness means that we should not have to rely on context to understand something.

We will see that our normal use of language depends on context all the time, as the same words can mean different things in different contexts, just as "quite" can mean "very" or "not much". In normal language people judge things not only by context but also relative to their own experiences; logical explanations need to be independent of personal experiences.

Understanding what is inherent in a situation involves understanding why things are happening, in a very fundamental sense. It is very related to asking "why?", repeatedly, like a small child, and not being satisfied with immediate and superficial answers. We have to be very clear what we are talking about in the first place. As we will see, logical arguments mostly come down to unpacking what things really mean, and in order to do that you have to understand what things mean very deeply. This can often seem like making an argument all about definitions. If you try having an argument about whether or not you exist, you'll probably find that the argument will quickly degenerate into an argument about what it means to "exist". I usually find that I might as well pick a definition that means I do exist, as that's a more useful answer than saying "Nope, I don't exist."

LOGIC AND LIFE

I have already asserted the fact that nothing in the world actually behaves according to logic. So how can we use logic in the world around us? Mathematical arguments and justifications are unambiguous and robust, but we can't use them to draw completely unambiguous conclusions about the world of humans. We can *try* to use logic to construct arguments about the real world, but no matter how unambiguously we build the argument, if we start with concepts that are ambiguous, there will be ambiguity in the result. We can use extremely secure building techniques, but if we use bricks made of polystyrene we'll never get a very strong building.

However, understanding mathematical logic helps us understand ambiguity and disagreement. It helps us understand where the disagreement is coming from. It helps us understand whether it comes from different use of logic, or different building blocks. If two people are disagreeing about healthcare they might be disagreeing about whether or not everyone should have healthcare, or they might be disagreeing about the best way to provide everyone with healthcare. Those are two quite different types of disagreement.

If they are disagreeing about the latter, they could be using different criteria to evaluate the healthcare systems, for example cost to the government, cost to the individuals, coverage, or outcomes. Perhaps in one system average premiums have gone up but more people have access to insurance. Or it could be that they are using the same criteria but judging the systems differently against those same criteria: one way to evaluate cost to individuals is to look at premiums, but another way is to look at the amount they actually have to pay out of their own pockets for any treatment. And even focusing on premiums there are different ways to evaluate those: means, medians, or looking at the cost to the poorest portion of society.

If two people disagree about how to solve a problem, they might be disagreeing about what counts as a solution, or they might agree on what counts as a solution but disagree about how to reach it. I believe that understanding logic helps us understand how to clear up disagreements, by first helping us understand where the root of the disagreement is.

In the first part of this book we are examining what logic is as a discipline for building arguments, and as a piece of mathematics. In the second part we'll see what the limitations of logic are. And in the third part we'll see how important it is, given those limitations, to take our emotions seriously.

LOGIC AS ILLUMINATION

Our aim throughout all of this is to shed light on the world. If we push our use of logic too far we risk overstepping that aim, and we will open ourselves to accusations of pedantry. Unfortunately mathematicians and extremely logical types of people are often accused of being pedantic by non-mathematicians or less logical people. Here, at the risk of sounding pedantic myself (and becoming very self-referential), I'm trying to shed light on the difference between pedantry and precision. I think the difference is *illumination*. I would characterize pedantry as precision that has gone further than necessary to shed light on a situation. There is plenty of precision that is there to clear things up, just like getting definitions right before trying to construct arguments with them. However, when extra precision does not help, I would call that pedantry.

Thus, self-referentially, I think that my distinction between pedantry and precision is itself a case of precision, not pedantry, because I think it sheds light on the situation.

Of course, people may disagree on where the lines are. One person's precision may well be another person's pedantry. It depends how much someone is seeking precision, and how much tolerance they have for ambiguity.

One of the difficulties that children have in learning about the world is dealing with the ambiguities of language. They are liable to take things rather literally because they haven't yet learned to use context to interpret the ambiguities. They haven't yet developed a tolerance for (or understanding of) subtle shades of nuance. A friend of mine recounted an incident when her small son was eating a bag of crisps and said he'd had enough. "You can just leave them on the table," she said, whereupon he obediently tipped them out onto the table.

As adults we develop the ability to become more relaxed about figurative language, and more relaxed about how precise we need things to be in order to get on with our lives. This is a bit like how accurately you need to measure things. When I'm weighing sugar to make a cake, I know it doesn't matter all that much if I'm off by 10g or so. However, when I'm weighing sugar to make macarons I know it matters an awful lot so I'll try to weigh to the nearest gram according to my digital scales. If someone is measuring the amount of anaesthetic to use to put someone under general anaesthetic I rather hope they'll measure extremely accurately. Indeed I was rather put out the one time I did go under general anaesthetic, for a knee operation, when the anaesthetist discovered I was a mathematician and said in that cheerful way people do, "Oh, I'm terrible at math!" I did not feel heartened.

I admit that I often find myself seeking more precision than some other people, and I do get accused of pedantry. But I'm convinced that I'm honestly just trying to shed light on situations. Actually, I tend to like more light in situations physically as well. I have bright lights on my desk, and I love bright sunshine because I like seeing everything more clearly. I like this in my thought processes as well. Achieving the illuminating precision sometimes takes longer – more thought, more words of explanation, more groundwork – and this is often unacceptable in today's world of soundbites, memes, and so-called mic-drops. It turns out that saying something with impact is often more important than saying something true. But there should be a way to show truth without sacrificing impact, and of having an impact without sacrificing truth. That is the best way to use logic in this complex world of unpredictable, emotional, beautiful humans.

I imagine shining a bright light to illuminate the things we are trying to understand. If we hold it close, the light will be bright but will illuminate only a small spot. If we move it further up, we illuminate a wider area but the light will be less bright. Eventually if we hold it too far out the light will become so diffuse that we won't see anything at all. But if we put it right on top of the things we're studying, we also won't see much.

Logic and abstraction are like shining a light at things. As we get more abstract, it's like raising the light off the ground. We see a broader context, but with less fierce detail; however, understanding the broad context helps us understand the detail later. In all cases the aim should be illumination of some kind. First we need some light, and then we can decide where, and how, to shine it.

2

WHAT LOGIC IS

DOES CHOCOLATE MAKE YOU HAPPY?

IF I EAT CHOCOLATE then I am happy. Is that logical?

If I touch wood after referring to something ominous, I feel better. If you fly from Chicago to Manchester via London, it can be cheaper than just flying to London on the very same flight and not going any further. If you drop some money in the street, someone will probably take it.

If you are white, then you have white privilege.

Are these things logical?

The innocuous little word "if" has a whole range of slightly different meanings. Some of them, but not all of them, encapsulate the most important building block of logical arguments: logical implication.

A logical argument is a way to demonstrate or verify that you are right. In life there are many ways to demonstrate that you are right. One is to shout loudly. Another is to tell anyone who disagrees with you that they are stupid. These are not good ways to persuade people that you are right, but unfortunately these ways are quite common.

The scientific method for demonstrating what is true involves carefully gathering evidence, analysing it, and then documenting the whole process in a way that could be reproduced by someone else if they followed the same steps. Importantly, it also comes with a way of discovering you are wrong.

Mathematics is at the heart of science but is a bit different from it. Mathematics uses logic, rather than evidence. It uses logic to decide when something is true, and it also uses logic to detect when something is wrong. We can sum this up by saying:

Logic is to mathematics as evidence is to science.

That is to say that the *role* of logic in mathematics is analogous to the *role* of evidence in science, but logic and evidence are fundamentally different. Unlike evidence, logic tells us when something *has* to be true, not by cause and effect, not by probability, not by observation, but by something inherent that will never ever change.

In this chapter we'll discuss the basic way that logical arguments are built: by logical implication. Logical implication is how you move from one true statement to another. It doesn't make more things true, it just uncovers more true things than we saw before. Logical implication says that "if" one thing is true "then" another must be true, using logic.

It becomes complicated because in normal life we say "if . . . then . . ." in situations that are not logical. It might be personal taste, like "if I eat chocolate then I am happy". It might be a threat, like "if you say that one more time then I will scream", or a bribe, like "if you eat your broccoli then you can have ice cream", or a promise, like "if you confide in me I won't tell anyone". It might be an agreement, like "if you walk my dog for me then I'll pay you twenty quid". It might be causation rather than logic, like "if you drop that glass then it will break". It might be rules, such as "if you are over 75 then you don't have to take your shoes off when going through airport security". It might be a personal opinion about standards of behavior: "If you loved me you wouldn't say that!" really means "I personally don't think that is a loving thing to say." Another problem is that we use "imply" to mean "insinuate" in normal language, as in "Are you implying I'm stupid?" We will explore these examples in this chapter, and think about the difference between these and truly logical implications. The difference is a bit blurry in normal life, but we can try and find the distinction by thinking about examples.

NORMAL LIFE EXAMPLES

Normal language is more vague than mathematical language, and so in normal language "if . . . then . . ." can signify some other things, as we mentioned above. So just because there's an "if . . . then . . ." in a sentence, it doesn't necessarily mean there's a logical implication.

The difference between formal logical language and informal real-world language is something that we will come across repeatedly as it is a source of a great deal of confusion, both about logic and about the real world. The primary aim of normal language is communication, whereas the primary aim of logical language is to eliminate ambiguity. These are not mutually exclusive aims. When we communicate, we try to do it as unambiguously as possible. And when we are trying to eliminate ambiguity, we are usually doing it to try and communicate more clearly. But normal language communicates with the help of context, body language, intonation, human understanding and so on. Logical language does not have the benefit – or confusion – of any of those things. "If . . . then . . ." can only mean one thing in logical language, but in normal life it depends on the situation.

All of these varied uses of "if . . . then . . ." are not exactly logical but they're not exactly illogical either: they don't *contradict* logic, they just aren't governed by it. The English language seems to lack a way of making this distinction, so we might say "non-logical" (although it's a bit clumsy) or "alogical", like apolitical, asexual or atheist. One of the points I will keep coming back to is that you can be alogical without being illogical, and indeed being alogical is unavoidable and sometimes beneficial or even crucial, whereas being illogical is undesirable.

So much for things that are not logical implications. What counts as logical?

"If you have white privilege then you have privilege."

This is a logical implication, because it comes from the inherent definitions: white privilege is one specific form of privilege. More contentious is this:

"If you are white then you have white privilege."

If we acknowledge that white privilege exists, then I do think this is logical. Perhaps in order to make it hold we need to be more specific about the context:

"If you are white in Europe or the US then you have white privilege."

or perhaps more obviously.

"If you are white in a place that has white privilege then you have white privilege."

You might think that the last deconstruction has become a bit pointless, and you have a point. The closer to a purely logical implication we get, the more obvious it should sound. This is the aim of finding the logic inside an argument: to make it more obvious.

However, although some people think this last statement is obvious, it is still contentious: some people claim that white privilege doesn't apply to *them*, only to richer white people. They are using a different definition of "white privilege". In Chapter 6 on relationships we will discuss the sense in which all white people have privilege, and the sense in which some white people are still lacking privilege for other reasons. The language itself is problematic as it is open to being used in so many different ways.

If we continue to use normal everyday language we are doomed to have problems being completely logical because the words we use are not completely logically defined, but we can get close enough that to call it anything other than logical would, in my opinion, be pedantry rather than precision. We'll now try finding the logical implications inside a different contentious argument.

SOCIAL SERVICES

Some people think social services should be expanded to give more help to vulnerable people. Others think social services should be cut to save money and stop encouraging laziness. Is there logic in either of these arguments? Does logic support one over the other?

One logical approach is to abstract these arguments down to the bare bones of false negatives and false positives. A false negative in this case is someone who deserves help but doesn't get it; a false positive would be someone who doesn't deserve help but does get it. Then the following implications become logical:

- If you care more about false negatives than false positives you will believe in expanding social services.
- If you care more about false positives than false negatives you will believe in reducing social services.

This is a simplification of the argument, but in performing this simplification we gain some clarity about the difference between those positions, and we see that a person who cares more about false negatives is simply never going to reach agreement with a person who cares more about false positives. In that situation the key would be to change someone's mind about that core principle rather than anything else.

False positives and false negatives turn out to be at the heart of many other disagreements. So in this case not only does the abstraction clarify the argument, but helps us make a connection with other arguments too.

For example, a much shared motivational mantra for life is

that you're "less likely to regret doing something and failing than not doing it and never knowing". This is supposed to encourage us that it's better to do things we perhaps shouldn't have done (false positive) than not do something we should have done (false negative). In fact, I rather like the line of the prayer that laments "We have left undone those things we ought to have done and we have done those things we ought not to have done": both false negatives and false positives.

This approach also helps me deal with jetlag: I have learned that I am better at staying awake when I'm tired (false positive) than going to sleep when I'm wide awake (false negative). Thus a better strategy for me to deal with jetlag is to undersleep in advance, knowing that when I arrive I can stay up when necessary and then be so tired at night time that I'm bound to fall asleep. Whereas some people are not so good at the false positive, so it's better to be well slept in advance, and then sleep even more upon arrival. The logical implications are these:

- If you're better at staying awake tired than falling asleep not tired you should deprive yourself of sleep in advance of crossing time zones.

- If you're better at falling asleep not tired than staying awake tired you should sleep well in advance of crossing time zones.

It might seem surprising that dealing with jetlag could have anything in common with an argument about social services, but this is one of the powerful aspects of abstraction, when it makes connections between apparently unrelated situations and thus makes more efficient use of our limited brain power. In Chapter 11 on axioms we will discuss how to use abstraction to uncover more of our own personal fundamental beliefs.

LOGIC AND DISCOVERY

If a statement follows from pure logic then it has to be true, automatically. Saying it out loud doesn't exactly add new information, but it does add new insight. This is why in normal language logical implications can sound a bit stupid, as the immediate new insight is often horribly obvious. Take the example "If you have white privilege then you have privilege." The "you have privilege" part of the sentence is the logical conclusion that automatically has to be true. It doesn't add any new information, but rather, it gives a different view point on the same thing; in this case the new view is the broader context of different types of privilege.

In this way logic is really about shedding new light on things rather than discovering new things. In a way, this is no different from, say, an archaeologist digging up an artifact. That artifact was already there, it's just that digging it up brings it to light. We get new insights but only because before that we were a bit ignorant. If we dig up a pot or the foundations of a building from hundreds of years ago, someone did already know about it, they just happen to be long dead.

Sometimes you might go on holiday abroad and "discover" a cute little café down some backstreet. Of course, you didn't actually discover something new – some other people already knew about it (the owners and all the people who already go there). But it's new to you. Sometimes people think they've "discovered" an amazing new singer, but they turn out to be extremely famous already, just not to the person who thinks they've just discovered them, and then everyone else rolls their eyes.

Logical conclusions are not new facts. Just like America was there all the time before white people reckon they "discovered" it, logical conclusions are true all along whether or not a human notices it. With "If you have white privilege then you have privilege" the conclusion is rather obvious, but the power of logic builds up when you string together a series of logical conclusions one after another, gradually getting you somewhere further from when you started. For example, we could string these implications together:

1. If you are white then you have white privilege.

2. If you have white privilege then you have privilege.

We now have the implication "If you are white then you have privilege."

Sometimes the revelation happens suddenly, like unearthing spadefuls of dirt until suddenly hitting treasure. Sometimes it happens gradually, like in the wonderful example of the guy who swapped a paper clip for a house.

WHEN BABY STEPS ADD UP

Kyle MacDonald is an internet legend who set himself the challenge of swapping a paper clip for a house. Not in one go, but by a series of swaps with people who didn't think they were getting a raw deal. It's true that the paper clip he started with was quite large (and red) – it wasn't just a bog standard office-supplies paper clip.

It sounds completely implausible, but he did it with this long series of trades; at each step someone thought the two things in question were equivalent enough to be traded but he got very far from his original paper clip in the process:

paper clip
↓
pen shaped like a fish
↓
hand-sculpted doorknob
↓
camping stove
↓
1000-watt generator
↓

"instant party" (beer keg with neon sign)
↓
snowmobile
↓
two-person trip to Yakh, British Columbia
↓
a large van
↓
a recording contract with Metalworks
↓
a room for a year in Phoenix, Arizona
↓
an afternoon with rock star Alice Cooper
↓
a motorized snow globe
↓
a part in a Corbin Bernsen film
↓
a house in Kipling, Saskatchewan

Apart from the mere fact of these trades having occurred, the thing that fascinates me about them is the question of why those trades were considered fair by the people involved. When MacDonald traded the afternoon with Alice Cooper for the snow globe, he had built up quite a following online, and this trade caused some consternation. But perhaps he knew what he was doing – did he know that the film director Corbin Bernsen collected these things, so he could make a good trade with him?

The final trade occurred because the people of Kipling (population: 1140) decided they wanted someone from their town to be in the film, so offered MacDonald a two-storey house in their town. He and his girlfriend moved in, apparently, in the autumn of 2006.

This episode fascinates me for many reasons, not least because it reminds me of a time when the internet wasn't overtaken by insults and bullying. But I am really fascinated by this mental version of an optical illusion, where you can take tiny steps that don't seem too surprising, and get somewhere that is extremely surprising and a very long way from where you started. This is how logic works. Each step you take is supposed to be entirely driven by logic, which means it should really just be an unpacking of some definitions, and should seem rather obvious and possibly even trivial. But when you stick them together in succession you can arrive somewhere that seems quite new, and very far from where you started. Kyle MacDonald's series of trades was virtuosic and masterful. Long chains of logical implications can also be virtuosic and masterful. They are how mathematics progresses, and I will later argue that I think they are an essential skill of a powerfully rational person.

Building up a long chain of implications to arrive somewhere new is the idea of a logical proof, and it is how logical proofs work in mathematics. Real life is not mathematics, but we should still try to make logical arguments in real life that work in a similar (though not exactly the same) way. Every step in the argument should be a logical implication.

A more serious example of a long chain of implications is a study into why babies were suffering birth defects, described in *The Power of Habit* by Charles Duhigg. The birth defects were found to be caused by malnutrition in the mothers. But this wasn't just during pregnancy – it was long-term malnutrition. Long-term malnutrition was found to be caused by poor nutrition, which was in turn caused by poor science education in school. That poor education was found to be caused by teachers not having a good enough science background in their training. So a surprising conclusion was drawn: requiring a higher level of science achievement in teachers would eventually reduce birth defects in babies. Incidentally Duhigg writes that the person who led this study was a young Paul O'Neill, who later became a renowned CEO and then United States Secretary of the Treasury. It is an example of a masterful construction of a long chain of implications in normal life.

IMPLICATION, FORMALLY

Even if our aim is to use better logical implication in normal life, I think it is important to understand some more of how it is used in mathematics. Mathematical language is dry and formal, which can make it seem offputting and irrelevant. But its dryness is there for the excellent reason of making things crisp and clear rather than soggy and mushy. It also helps make things more concise, which in turn helps us build bigger and more complex arguments. It's a bit like those vacuum bags into which you put your clothes and then suck all the air out with a vacuum cleaner, condensing a whole pile of clothes into one compact unit.

A more concise way of saying "If ... then ..." is "implies". So instead of saying "If A then B" we can say "A implies B". Mathematicians use the symbol $\implies$ for "implies". This implication means that whenever A is true, B absolutely has to be true. When A is false, the implication doesn't tell us anything.

For example, "Being a US citizen implies you can legally live in the US" tells us that whenever someone is a US citizen they can legally live in the US. But when someone *isn't* a US citizen this implication doesn't tell us anything about them one way or another: they might be able to live in the US legally (for example, if they have a visa or have permanent residency) or they might not. Unfortunately this logic is lost on some people who think that anyone who isn't a citizen is illegal. We will come back to this grave error of logic in the next chapter.

A proof is basically a whole series of implications strung together like this:

$$A \implies B$$
$$B \implies C$$
$$C \implies D$$

We can then conclude that $A \implies D$. This is because if A is true then B is true by the first implication, and next if B is true then C

is true by the second implication, and finally if C is true then D is true by the third implication, so the overall knock-on effect is that if A is true then (after some thought) D is true.

The critical point is "after some thought" – a chain of implications takes more concentration and command of logic to follow than a single one does, and these resources are unfortunately often lacking from arguments.

Here are some longer chains of implications that take more than a basic command of logic to follow them:

1. If you say women are inferior, that is insulting to women.
2. If you think that "feminine" is an insulting way to describe a man, you are saying that women are inferior.
3. Therefore if you think that "feminine" is an insulting way to describe a man, you are insulting women.

Here is another:

1. If you don't stand up for minorities being harassed then you are letting bigotry flourish.
2. If you let bigotry flourish you are complicit with bigotry.
3. If you are complicit with something bad then you are almost as bad as it.
4. Therefore if you don't stand up for minorities being harassed you are almost as bad as a bigot.

It is important to note that the conclusion is true if you don't stand up for minorities, but the *implication* is true whether or not you stand up for minorities. I might not know that you are a great ally of minorities, and I might say to you "If you don't stand up for minorities who are being harassed then you are almost as bad as a bigot yourself". My statement is still true although you personally are not actually anywhere close to being a bigot. This is an important subtlety of implication. "If you are a US citizen or permanent resident you are required to have health insurance" is true whether or not you are in fact a citizen or resident. The implication doesn't tell us whether or not someone needs health insurance; we only know they do if we already know they are a citizen or resident. (This point was lost on one insurance agent who told me everyone in the world needed US health insurance regardless of whether they even lived there or not.)

We can build up huge arguments from logical implications, and the argument will still have this feature: the concluding statement is only known to be true when the opening conditions are fulfilled. But the argument itself tells us that *if* the opening conditions are fulfilled *then* the concluding statment is true, and this argument is always correct. This sort of justification, in mathematics, is a proof.

I am gradually going to argue that stringing together long chains of implications gives us logical power. It is what enables us, like Kyle MacDonald, to start from something obvious and work our way to something complex and unobvious. Being able to construct and follow such complex arguments is hard but is a critical part of making good use of our human brains. I believe it is one of the things that separates us from simpler animals and tiny children, who can only deal with immediate needs and direct observations. Long chains of implications often require us to package many connected ideas into a single unit so that we can build on them more easily, like vacuum packing our clothes. What we gain in the process is new insights and deeper understanding.

WHAT DOES A PROOF LOOK LIKE?

Before we start a proof in mathematics we have to lay the groundwork very carefully, like setting up the rules for a sport. There are several aspects of this, and I think it's illuminating to think about them because many arguments in real life go wrong because of problems with the groundwork rather than with the argument per se. Often in arguments in real life we only realize we're using different definitions or assumptions at a rather late stage of the argument.

1. We should carefully define the concepts we're talking about.
2. We should carefully state the assumptions we're making.
3. We should carefully state exactly what we're going to prove, in an unambiguous way.

Assumptions in mathematics are a bit different from assumptions in life. In math it is to do with conditions under which we've decided to work, or conditions under which we think our result is true. Then when we apply the result, we first check whether those conditions are fulfilled in the situation we're trying to apply it to.

For example, we might assume we're living on the surface of a sphere, and then show that something is true under those conditions. This does not pass any judgement about whether or not we actually are living on the surface of a sphere. It just says that if we do turn out to be living on the surface of a sphere, then this thing will be true.

Assumptions in life should work similarly but unfortunately often don't. For example, in arguments about why women earn less than men, on average, sometimes people assume that women don't care about earning money as much as men do. Now, under that assumption it might well be reasonable that they end up earning less than men. But that is a hypothetical world. When we apply that result to the real world, we need to see if our assumption is true: is it really the case that women don't care about earning money as much as men do?

Note that I don't actually think the implication is true: even if it were true that women didn't care about earning money as much as men do, I don't think that would make it reasonable to pay them less than men for the same job. I think that is exploitation. In any case, if we're clear about what our assumptions are, we can at least be clear about what aspect of the argument we're agreeing or disagreeing with.

Once we've laid the groundwork, the actual proof consists of a series of statements, each of which follows logically from what is already known to be true. This could either be the assumptions, known truths about the world we're in, or previous statements in the proof. The series of statements creates a chain of logic from the starting point, which is the assumptions we're making, to the ending point, which is the thing we're aiming to prove. Of course, sometimes this chain breaks down, especially in normal life, which is why there are more bad arguments in the world than good ones. These failures separate broadly into problems of knowledge and problems of logic.

PROBLEMS OF KNOWLEDGE

When an argument breaks down because of problems of knowledge it might be in one of these ways:

- Unstated assumptions, or using stated assumptions incorrectly.
- Incorrect definitions, or incorrect use of definitions.

Using an unstated assumption is one way of covering up the fact that you don't know something. Using an incorrect definition, or using a definition incorrectly, is a way of making an argument easier than it should be, leading to an abundance of straw man and false equivalence arguments.

These ways that a mathematical proof goes wrong are often ways that arguments in real life go wrong as well. Unstated assumptions often occur in arguments about benefits, when some people tacitly assume that people are only poor if they are too lazy to work hard. Or in arguments about abortion when some people assume that unwanted pregnancies only occur if people are promiscuous. Or in arguments about clinical depression when people assume that depression is caused by circumstances and therefore there is no reason a successful person should be depressed.

Problems of incorrect definitions often occur in arguments about immigration, when some people take the definition of "immigrant" to be "illegal immigrant". In normal life the problems often occur because in fact no definition has been stated. This often happens in arguments about whether some behavior is "patriotic" or not, or whether something is "sexist", or whether someone is or isn't a "feminist". Or whether something is or isn't "democratic".

PROBLEMS OF LOGIC

Problems of logic in proofs include:

- Gaps in the logic: leaping from one statement to another without justifying it, or leaving out too many steps in between.

- Incorrect inferences: this is when a logical step is made that is actually incorrect, where you say something follows logically from something else, but it doesn't.

- Handwaving: arriving at a conclusion without true use of logic, but by metaphorically waving your hands around enough that people think you are.

- Incorrect logic: there are many subtle ways that incorrect logic can get slipped into arguments as logical fallacies and we will examine some popular ones in some detail in the second half of the book.

Handwaving often occurs backed up by yelling and insults in arguments in normal life. Comments such as "Anyone with half a brain cell can see this!" is usually a sign that someone doesn't actually know how to justify something.

An example of an incorrect inference is the view that "scientists agree with each other, which shows there is a conspiracy". The conclusion (there is a conspiracy) does not follow logically from the fact that people are agreeing with each other, as shown by my earlier counterexample of the Wimbledon result.

An example of a gap in logic is when one person is blamed for something, and all the other factors are ignored. Consider a so-called "swatting" tragedy when a hoax call is made to the police who then storm an innocent person's home and shoot them dead. If the police often place all the blame on the hoax caller they are leaving out the fact that they shot an innocent person dead with precious little evidence or reason.

An argument using incorrect logic should be straighforward to refute, but only if you have a good understanding of logic *and* of the emotions behind the faulty logic. We will come back to this at the end of the book.

WHERE DO IMPLICATIONS START?

There is a story that Stephen Hawking tells in *A Brief History of Time*, about an audience member who came up to a "well-known scientist" after a talk on cosmology, and declared "What you have told us is rubbish. The world is really a flat plate supported on the back of a giant tortoise." The scientist asks what that tortoise is standing on, whereupon the audience member supposedly replies, "You're very clever, young man, very clever, but it's turtles all the way down!"

Even if we are using logic instead of turtles to hold our arguments up, we still need to wonder what is holding up each level of our argument. It is important to remember that logical implication only enables us to deduce something from something else. "X implies Y" only tells us that Y is true *if* X is true. It doesn't tell us whether or not X is true. In order to know that X is true we need something to imply it, perhaps "W implies X". But what tells us that W is true? Perhaps V implies W. But what implies V? Where does all this originate from?

As I have mentioned, I think this process of working backwards is just like small children who ask "Why?" repeatedly. Every time you give them an answer they proceed to ask "Why?" to that. Children have apparently infinite curiosity and infinite tolerance for doing the same thing over and over again, so are likely to keep asking "Why?" until the adult gets frustrated and puts a stop to it. I personally carried on asking why all the way through science and mathematics until I was doing my own research in abstract mathematics because I was still asking "Why?". Pragmatism and worldly responsibilities cause adults (most of them anyway) to stop asking why, accept some things, and get on with daily life. I like to think that we still all have that infinite curiosity inside us, which is why Wikipedia is so popular and why so many of us are prone to falling down Wikipedia wormholes. You know the kind, where you keep clicking on more links to read more articles and understand more things. (I have done my fair share of cat video wormholes too.)

WHERE DO WE STOP?

Knowing where we start is important. But in a way this comes down to knowing when to stop, that is, knowing when to stop asking why and stop trying to justify ourselves further. When should we stop filling in gaps in the logic? Eventually we have to close Wikipedia and get on with something else. Eventually we have to tell the inquisitive child that it's time to go to bed, or get ready for school. If I imagine an adult who has not developed the ability to stop asking those "why" questions and get on with daily life, I imagine a tortured philosopher, endlessly questioning everything and searching for meaning instead of eating, sleeping or earning money. Do I exist? What is the meaning of life? Why are we here? Why is there so much suffering in the world? What is love? Why do people hate each other? Why do people hurt each other? To be a moderately functioning adult, we have to stop asking those questions at some point – not necessarily permanently, but at least for some part of every day.

In logic we also have to stop asking questions at some point, and accept some facts, otherwise we'll never get anywhere. We can work out that Y is implied by X, which is implied by W, which is implied by V, and so on, but at some point we need to stop working backwards and decide that we have explained enough for now. The place where you stop trying to work backwards is the place where you have basic assumptions or beliefs that you don't try to justify at the moment. It doesn't mean that you won't ever try, and that nobody else will, it's just that right now you've decided this is your starting point and is the basis of your logical system, or your belief system.

In logic and mathematics these are called axioms, and we will discuss them further in Chapter 11. We need a starting point in logic because we can only deduce things from other things – we can't deduce something from nothing. We're not magicians, and even magicians don't really produce something from nothing, they're just very successfully fooling us. If a logician claims to be deducing something from nothing they're fooling us too.

Looking for axioms, or starting points, in my own system of beliefs has led me to a much clearer understanding of my own thinking. It has enabled me to identify basic beliefs that some other people don't hold. For example, when it comes to prejudice, I fundamentally believe that prejudice of those with more power towards those with less power is much more harmful than the other way round. This means that if I can trace someone else's axioms down to different ones then I know we have a disagreement at the very beginning of the argument, and there's no point trying to resolve a later implication without trying to resolve this starting point. But crucially it is possible to reach a different conclusion from me by applying logic perfectly soundly but starting from different basic beliefs. So two people can both be logical but still disagree about things.

Working out what the basic starting points in an argument are is an important part of analysing it logically, and is an important part of understanding the nature of disagreements. The logic must all flow out of that starting point. In the next chapter we'll talk about the direction of that flow. Like time, logic has a direction and we must not try to violate it.

3

THE DIRECTIONALITY OF LOGIC

DOES BEING HAPPY MAKE YOU EAT CHOCOLATE?

EATING CHOCOLATE MAKES ME HAPPY instantly. It has to be good chocolate, but it works without fail every time.

Does being happy make me eat chocolate? That's a completely different question.

On a more serious note, being a US citizen means that you can legally live in the US. If you can legally live in the US, does that necessarily mean you are a US citizen? This is a completely different question. Some people erroneously think that being a citizen is the only way to be legally resident, but there are plenty of other ways, including having a work visa, permanent residency, or being admitted as a refugee.

Time and causation flow in one direction only, and so does logic, and we must be careful not to make errors in direction. In the last chapter we described the string of swaps Kyle MacDonald made, starting with a paper clip and ending up with a house. I sometimes wonder if these swaps were reversible. If he changed his mind about, say, the snow globe, would the original owner have taken the item back? It's not clear.

In the example in the previous chapter, I argued that if you don't stand up for minorities who are being harassed then you are almost as bad as an outright bigot. What if we turn this round? If you are almost as bad as an outright bigot does that mean you don't stand up for minorities who are being harassed? No, there are plenty of other ways to be "almost as bad" as a bigot even if you do stand up for harassed minorities and think that this completely exonerates you. Perhaps you stand up for them in public but then quietly block their promotions or pay rises, turn them down for jobs, or refuse to vote for them.

The point is that we have an implication like this:

$$\begin{matrix} \text{not standing up for} \\ \text{harassed minorities} \end{matrix} \implies \begin{matrix} \text{almost as bad} \\ \text{as a bigot} \end{matrix}$$

but we can't just reverse the arrow to get this:

$$\begin{matrix} \text{almost as bad} \\ \text{as a bigot} \end{matrix} \implies \begin{matrix} \text{not standing up for} \\ \text{harassed minorities} \end{matrix}$$

The fact that the implication sign $\implies$ looks like an arrow is not a coincidence. It is chosen to help us see that the logic flows in one direction only. Turning the arrow around would change the meaning, possibly drastically. Take the logical privilege example from the previous chapter. In the original form it was

$$\text{you have white privilege} \implies \text{you have privilege}$$

If we turn the direction of the arrow around this becomes

$$\text{you have privilege} \implies \text{you have white privilege}$$

This is obviously not true as there are plenty of other types of privilege that you might have even if you're not white, such as the privilege of being born to rich or powerful parents.

What about this:

$$\text{you are a woman} \implies \text{you have experienced sexism}$$

This is the premiss of the Everyday Sexism project: that *every* woman experiences sexism, even if it's not overt. It might take the form of micro-aggressions that we're supposed to brush off, and perhaps we're so accustomed to it that we barely even register it any more. The sad fact that we take it for granted as a part of life doesn't mean it's not there – on the contrary it means that it is everywhere.

Now let's try turning the arrow around:

$$\text{you have experienced sexism} \implies \text{you are a woman}$$

This is a completely different question, but unfortunately often gets muddled up with the first one. If you say "all women experience sexism" someone (usually a man) is likely to protest that men experience sexism too. This may or may not be true, but in any case it is not logically related to the first question. The first implication says that *if* you are a woman *then* you have experienced sexism; it does not claim anything if you are a man.[1]

All these examples show that the act of turning the arrow around in an implication statement makes a completely new statement for us to think about. The statement that we get by turning the arrow around is called the *converse* of the original statement.

ON BROCCOLI AND ICE CREAM

One of my favorite examples of implications and converses is "If you eat your broccoli you can have ice cream." I will admit up front that this gets quite confusing in normal language, which is why mathematicians quickly prefer using letters and symbols to keep things clear. But let's try with words.

A logical, literal child might start asking what other foods will imply ice cream, if they really want to avoid broccoli. Here their precision comes across as pedantry to the adult, who might say exasperatedly "You know what I mean!", but the child is just seeking clarity and trying to find a loophole to avoid

---

[1] The question of whether or not men experience sexism comes down to the definition of sexism. According to some theories of prejudice, sexism and racism should only be used for instances that fit into a bigger picture of systematic oppression. The idea is that when an oppressed group is biased against their oppressors, this is not the same as when the oppressors are biased against the oppressed group as a form of control. Whether or not we agree with these definitions I think it is important to notice the difference between oppressed people and oppressors, and we will come back to this in Chapter 13 on analogies. Thinking of it in these abstract terms clears up something about the difference; giving it a name would help us think about it.

eating broccoli. (I was not that child – I have always loved broccoli. Perhaps because it was never used as a threat to me. Or maybe it was never used as a threat because I loved it.)

The child might say "How about if I eat some fish instead?" to which the answer might be "No, you *have* to eat your broccoli!" or "No, you get ice cream *only if* you eat your broccoli!" These are both examples of a converse, but it's a bit hard to see with the broccoli example because it's not really a logical implication – more of a bribe.

Here are the two statements the parent made. First they said

If you eat your broccoli you can have ice cream.

broccoli $\implies$ ice cream

which guarantees the *child* that broccoli leads to ice cream. It says that if the child eats broccoli that is *sufficient* for earning ice cream.

The parent then said

You can eat ice cream only if you eat broccoli.

ice cream $\implies$ broccoli

which guarantees the *parent* that the child can't sit and try to find other ways of earning ice cream. It says that broccoli is *necessary* for earning ice cream, and there is no way round it. It is the converse of the first statement. (If the direction of this arrow looks strange, you can think of it as saying that if we subsequently see the child eating ice cream we can logically deduce that they must have eaten their broccoli.)

All this is to explain why "only if" is a way of expressing the converse of "if" – logic is flowing in the opposite direction. To ensure both the guarantee to the child and the guarantee to the parent, the promise technically needs to be "You can have ice cream *if and only if* you eat your broccoli". The trouble is that probably only a rather pedantic mathematician would bother saying that, so we grow up with the vague sense that "only if" means the same as "if and only if". Making the distinction in normal language is probably pedantic because it oversteps the goal of clarification. The trouble is that not making the distinction causes confusion for people when they do start thinking about logic formally. This can lead to much more serious consequences in more serious situations.

Imagine you're trying to catch a group of bank robbers and you know the whole gang was white men. So you know

> If someone you encounter is in that gang
> then they are a white man.

This is equivalent to

> Someone you encounter can only be in that gang
> if they are a white man.

So we can start by looking for white men. But finding a white man does not ensure that we have found a criminal, because the converse isn't true. The converse would be

> If someone you encounter is a white man
> then they are in the gang.

Being a white man is a necessary condition for being in the gang, but not sufficient.[2]

All this is very prone to get confusing and make your head spin in normal language, which is one of the reasons mathem-

---

[2] This approach of looking for white men would make sense if you knew that the gang had five people and you also know there are exactly five white men in the country. It might even be reasonable if you know there are exactly ten white men in the country, because if you round up five of them the odds are quite high that you'll get some of the actual criminals. This approach becomes gradually less like logic and more like racism as the country's population of white men grows. In reality this profiling is likely to be done against non-white people, not white people. How big does the country or city's population of the minority have to be before this veers from rational into racist? This is a question of gray areas that we will come back to in Chapter 12, and we have to be very careful that incremental arguments do not give us logical justifications of prejudice.

aticians reduce things to letters and symbols, because it can be easier to see patterns. Using arrows we have:

- True: gang $\Longrightarrow$ white
- False: white $\Longrightarrow$ gang

We'll deal with falsehood in the next chapter.

USING ARROWS TO HELP US

Mathematical notation is one of the things that can make mathematics baffling and difficult to learn. However, the notation is there to help us think clearly. Implications and converses demonstrate this. This can be extra confusing in normal language, because of the flexibility of English grammar and where we can put the word "if" in the sentence.

You can have ice cream if you eat your broccoli.

is logically the same as saying

If you eat your broccoli you can have ice cream.

In general we see that "if A is true then B is true" means the same as "B is true if A is true". This might look like we've turned around the flow of logic, but we've actually only turned around the grammar.

One advantage of using arrow notation is that the flow of logic is completely clear from the direction of the arrow.

The converse of

$$A \Longrightarrow B$$

is

$$B \Longrightarrow A$$

But also

$$A \implies B$$

is equivalent to

$$B \impliedby A$$

as it doesn't matter which way the arrow is facing on the page, only where it is pointing from and to. Indeed, we would mean the same logically (although maybe it would be a bit questionable emotionally) if we drew it like this:

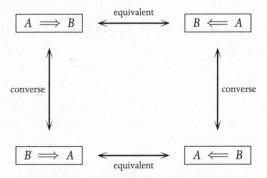

So we have these possibilities:

$$
\begin{array}{ccc}
\boxed{A \implies B} & \xleftrightarrow{\text{equivalent}} & \boxed{B \impliedby A} \\
\Big\updownarrow \text{\scriptsize converse} & & \Big\updownarrow \text{\scriptsize converse} \\
\boxed{B \implies A} & \xleftrightarrow[\text{equivalent}]{} & \boxed{A \impliedby B}
\end{array}
$$

If we try this using if/then again, the converse of "if A then B" is "if B then A".

The new statement looks superficially similar to the old one but is logically completely different.

USING VENN DIAGRAMS TO HELP US

Venn diagrams can help us picture some aspects of logic. I find pictures crucial when I'm doing math research. I often look like I'm staring into space, but what I'm really doing is manipulating pictures in my mind. Math gets its power from being abstract, that is, removed from the real world of objects and things we can touch. The trouble is that means it's hard to get a feel for it. One thing that helps is having pictures that capture some aspect of what you're thinking about. The pictures are like analogies (with apologies for the meta-analogy) – they don't *exactly* represent what you're thinking about, but they sum up some important aspect of it. They help us make the transition between the dry logic and our feelings. Tristan Needham said in his book *Visual Complex Analysis*:

> While it often takes more imagination and effort to find a picture than to do a calculation, the picture will always reward you by bringing you nearer to the Truth.

I think that is putting it a bit too strongly – some people really prefer symbols and words to pictures. But I find pictures very helpful. Venn diagrams are quite helpful for basic situations, and in Chapter 5 (on blame and responsibility) we'll see that when things get really complicated flow diagrams are better because they have more possibilities. Venn diagrams aren't so useful if you have more than about three sets because they get too chopped up for our eyes to take in.

In the first instance Venn diagrams can help us to see the directionality of implications.

Let's think about this logical implication:

If you are from England then you are from the UK.

We could draw England inside the UK abstractly like this:

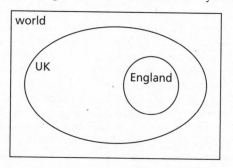

This sounds obvious but some people get upset about me saying I'm from England, because I don't "look English". But they don't mind me saying I'm from the UK. Logically they must think I am in the part of the Venn diagram that is inside the UK but outside England. However, I'm not from Scotland, Wales or Northern Ireland either.[3]

We could draw a diagram like this for any statement of implication even if it's not geographical and doesn't depict physical positions. For example:

If you have white privilege then you have privilege.

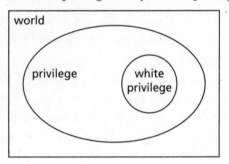

---
[3] I recently discovered that, technically, diagrams like the ones above aren't supposed to be called "Venn diagrams" but rather "Euler diagrams". To qualify as a Venn diagram, it is apparently supposed to exhibit all possible logical combinations of the sets in question. The ones above don't qualify because one circle is completely contained within the other, so there isn't a region where you can be in the small circle without being in the big one. Of course, this is the entire point of that particular diagram. Personally I think the distinction between Venn diagrams and Euler diagrams is more like pedantry than precision, so I shall go on referring to these as Venn diagrams, especially as that term is much more well known.

If you are a US citizen then you can live in the US legally.

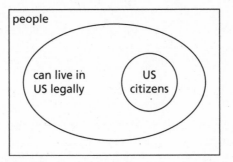

For the general case we have this:

$$A \implies B$$

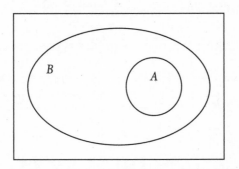

It is now a bit more vague what these circles are representing, so it is more of a schematic diagram at the moment than a rigorous one. But to me it does capture the idea that *A* somehow can't escape being part of *B*.

The Venn diagram also makes it visually evident that the implication does not go backwards automatically, because the inner circle and the outer circle (oval) are really playing different roles. *A* can't escape being part of *B*, but *B* can escape being part of *A*, because there is space in *B* around the outside of *A*. This corresponds to the logical fact that even if *A* implies *B*, it is still possible for *B* to be true when *A* is false. The following version is mathematically correct but misleading.

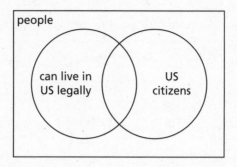

because it looks like there's a way to be a US citizen without being allowed to live in the US legally, as well as there being ways to live there legally without being a citizen. In fact, the real situation is not symmetrical like that –the right-hand area is empty. Logical implication is not symmetrical.

OUR WONDERFULLY AND CONFUSINGLY FLEXIBLE LANGUAGE

We have seen plentiful different ways of saying the same logical implication in words. Here is a full list of different ways of saying in words

$$A \implies B$$

including phrases putting A first and partner phrases putting B first:

- A implies B.
  B is implied by A.
- If A then B.
  B if A.
- A is a sufficient condition for B.
  B is a necessary condition for A.
- A is true only if B is true.
  Only if B is true is A true.

I think it is hard to be convinced that all eight of these statements mean the same thing, and I expect some people will write to me telling me I've got it wrong (so I'm checking rather carefully that I don't have any typos). I think the last is the most confusing. Here is a tragic example.

In a chilling recent incident, a policeman in Cobb County, Georgia was caught on dashboard camera reassuring a panicky white woman that "we only shoot black people". This is logically equivalent to saying

We shoot you only if you are black.

which is in turn equivalent to

Only if you are black do we shoot you.

or in arrows

you are black $\Longleftarrow$ we shoot you

and equivalently

we shoot you $\Longrightarrow$ you are black

or in words

If we shoot you then you must be black.

This is the reason that when you hear a report of someone being shot in a traffic stop in the US, you might feel fairly certain they were black.[4]

This is one reason that I prefer using symbols: it is quicker, clearer to me, and all eight of these phrases become the same, so I don't have to use up precious brain cells thinking about what things mean.

---

[4] Note that the logical equivalence of these statements doesn't mean they are true and it certainly doesn't mean they are morally justifiable. It just means that logically they mean the same thing. White people are killed by police too, including in traffic stops. The questions of whether black people are killed disproportionately and whether this is because of racism are different, much more complex questions.

CONVERSE ERRORS

Converse errors occur when someone makes the mistake of thinking the converse of a statement is equivalent to the statement. It's an understandable mistake to make in a way, given that there are eight ways of saying "*A* implies *B*" and eight ways of saying the converse. This happens when you tell students that they have to work hard in order to do well, and then they think that if they work hard they should automatically do well. Working hard is a necessary but not sufficient condition for doing well. It is not sufficient because you also have to work hard in the right sort of way, and if you think otherwise you are making a converse error.

In fact, the converse of a statement is logically independent of the old statement, which means that there is no logical connection between the two. That is to say just because one of them is true, it does not necessarily mean the other one is true, nor does it necessarily mean the other one is false. In fact, all combinations of true and false are possible, as shown by the following examples:

1. If you are a US citizen then you can legally live in the US. This is a true logical implication. The converse is "if you can legally live in the US then you are a US citizen". This is not true, as you can legally live in the US by being a permanent resident or having a visa.

2. If you have a university degree then you are intelligent. I do not believe this is true. I think there are some degrees awarded to people who are not intelligent; sadly the pass mark is very low. The converse is "if you are intelligent then you have a university degree". This is also not true – I think there are intelligent people who do not have a university degree, especially in older generations when going to university was not such a standard next step in life.

3. If you have experienced prejudice then you are a woman. This is not true; men and gender non-binary people can experience prejudice; gender non-binary people certainly do. The converse is "if you are a woman then you have experienced prejudice". I think this is true, whether or not you have ever acknowledged or complained about it.

4. If you support Obamacare then you support the Affordable Care Act. This is true because Obamacare is simply an informal name for the Affordable Care Act. This means that the converse is also true, "If you support the Affordable Care Act then you support Obamacare". Unfortunately there are people who support ACA but refuse to support Obamacare, not realizing that these are the same thing. They have such a strong objection to anything to do with Obama that rebranding something with his name in it was enough to put them off. This is eye-opening, and I think we can learn important lessons from it to do with how the way we present things really does matter and can override even very clear-cut logic.

We can sum up those conclusions in this table:

	original	converse
statement 1	true	false
statement 2	false	false
statement 3	false	true
statement 4	true	true

This is all four possible combinations of true and false for the statement and its converse. This means that if we start with a new statement, discovering it is true or false doesn't help us know anything about the converse, because it could still in theory be either true or false.

LOGICAL EQUIVALENCE

We have talked about the error of mixing up a statement and its converse, and the error of thinking that just because a statement is true its converse must also be true. However, sometimes both a statement and its converse happen to be true. In this case we have a situation of logical equivalence. That is, if A implies B and also B implies A, then A and B are logically equivalent – whenever A is true, B must be true, and also whenever A is false, B must be false.

This means that A and B are logically interchangeable, and are usually just different viewpoints on the same thing. Crucially, this doesn't mean they're exactly the same, as illustrated in the example of Obamacare and ACA above. Logically those are the same, but emotionally they are very different to some people, who feel fine supporting something with the calm and compassionate name "Affordable Care Act", but can't bear the idea of supporting something referring to Obama. Some other people have the opposite response, where the reference to Obama makes them *more* positively disposed towards it. We will later come back to the logical fallacy of false equivalence, where two things are assumed to be logically equivalent when they are not, such as when having a degree is taken as being equivalent to being intelligent. However, the Obama/ACA example is a case of "false inequivalence", where some people take things to be different, when in fact they're logically equivalent. Still, we should accept that they are not *emotionally* equivalent, and work with this fact rather than simply deny it on the grounds that it contradicts logic. We will come back to these issues in Chapter 15 on emotions.

When two things are logically equivalent, the implication flows in both directions, so we use this symbol: $A \iff B$. In words there are several ways of saying this, and they come in symmetrical pairs as the logic flows both ways:

- *A* is true if and only if *B* is true.
  B is true if and only if *A* is true.
- *A* is a necessary and sufficient condition for *B*.
  B is a necessary and sufficient condition for *A*.
- *A* is logically equivalent to *B*.
  B is logically equivalent to *A*.
- If *A* is true *B* is true, and if *A* is false *B* is false.
  If *B* is true *A* is true, and if *B* is false *A* is false.

This last pair sheds some light on the fact that "*A* implies *B*" tells us nothing about the case when *A* is false: if we want to deduce something from *A* being false we need the converse although, ·on the face of it, the converse gives us a way to deduce something from *B* being true rather than from *A* being false. We'll come back to this in the next chapter, in which we are going to explore what it means for things to be false.

4

OPPOSITES AND FALSEHOODS

HOW WE ARGUE AGAINST THINGS

THERE ARE ONLY TWO DEBATES I remember from the debating club at school. One was "This house believes that Margaret Thatcher should go", which was particularly memorable because she actually resigned the morning of the debate. The other one was "This house believes that strawberries are better than raspberries", a classic pointless and contentless debating topic. It is easy to think that both sides of this debate have an equally impossible task, because how could you ever argue that one kind of berry is better than another? What does "better" mean? However, the key with this type of debate is that the house only decides whether they will support the motion or not. So the proposers have to argue that strawberries are better than raspberries, but the other side only has to argue that the proposers are wrong. There are many ways they could be wrong. One way would be if, in fact, raspberries are better. But they could also be wrong if strawberries and raspberries are equally good. Or if "better" is impossible to define. Or if the entire idea is idiotic.

Most arguments are not like debates, but still consist of someone claiming that something is true, and the other person saying they're wrong. If they are attempting to be logical, the first person will try to justify what they're saying by constructing a logical argument to back it up. The second person should then either try to find a flaw in their logical argument, or try to construct their own logical argument to back up their assertion that the person is wrong.

Logic, mathematics and science are all ways of finding out

what is true. But they are also ways of finding out what is not true. Admitting the possibility of being wrong and having ways to detect it is an important part of being a rational human being, I am sure. (But I could be wrong.)

Negation is how we argue against things. Unfortunately in normal life we often do it very badly. Arguments degenerate into insults, intimidation or yelling all too quickly, especially in online comments sections. I optimistically think that this is out of frustration at one's own failure to get a point across, rather than because everyone enjoys insulting, intimidating or yelling at people. My optimism leads me to do things like write books about logic, because I think most people can do better than that, and I even think that most people might want to. Or failing that, could be persuaded to want to.

One reason that arguing against things in normal life doesn't go very smoothly is that we don't always properly understand which things are equivalent to each other and therefore which things are true refutations. Once we understand negation, we can start to build logical power; one of the first steps is to understand many different points of view on the same idea, and how they agree or oppose each other.

NEGATION VS OPPOSITE

Imagine a debate about education systems in which someone claims, as someone periodically does, that the Asian education system is better than the British or American one. There are two ways to oppose this view:

1. Measured and calm: I don't think the Asian education system is better.

2. Extreme and excited: No way! The British education system is better!

Emotional tone aside, these are logically two different ways of opposing the original view. The second (extreme) way is what we think of as being the "opposite" in normal language. But it is not the *logical* opposite. In logic, to make the negation we take the original statement and simply declare that it is not true. This is the first (calm) opposition above: the negation of "the Asian education system is better than the British one" is "It is not true that the Asian education system is better than the British one". Or, to put it in more natural wording, "The Asian education system is not better than the British one." Just like the question of strawberries and raspberries, there are many ways in which the Asian education system could be "not better". What does "better" mean, in any case? What are the systems aiming to do? How are we measuring what they achieve? What do we want an education system to achieve? Some people seem to measure everything in terms of math and science achievement, or other standardized test results, whereas other people want to measure in terms of readiness for the workplace. Is that all we want education to do – train people to score highly in standardized tests and be a good employee?

Here are some more examples showing the difference between logical negation and "opposites" in normal language:

- *Original statement:* I think the EU is fantastic.
  Opposite: I think the EU is terrible.
  Negation: I do not think the EU is fantastic. This is not the same as thinking it's terrible. It is possible to be "not fantastic" without being "terrible". For example, it could be mostly very good but with some flaws. However there is a tricky issue of language here because with a particular tone of delivery it could sound like a wry understated way of saying it's terrible. However, this is a quirk of language rather than a logical negation.

- *Original statement:* Margaret Thatcher was the greatest Prime Minister.
  Opposite: Margaret Thatcher was the worst Prime Minister.
  Negation: Margaret Thatcher was not the greatest Prime

Minister. But perhaps she was not the worst either – she could have been the second worst, or the tenth worst, or something.

- *Original statement:* Climate change is definitely real.
  Opposite: Climate change is definitely fake.
  Negation: Climate change is not definitely real. Is anything definite about anything? However, that doesn't mean it's definitely fake – it's highly likely to be real because of a huge quantity of evidence pointing towards it, where here "real" means a scientifically sound theory according to the strict framework of science.

- *Original statement:* Sugar is good for you.
  Opposite: Sugar is bad for you.
  Negation: Sugar is not good for you. But it's also not directly bad for you as a small amount every day probably won't do you any harm, it's just that in large quantities it's probably bad for you.

- *Original statement:* I am male.
  Opposite: I am female.
  Negation: I am not male. I still might not be female as I could be one of the estimated 1.7 percent of the population born intersex.

In general the negation is a broader statement than the opposite. The opposite is the exact opposite extreme, or, as we might say for emphasis, the "polar opposite". The polar opposite of the North Pole is the South Pole, but there's an awful lot of world in between those poles.

- *Original statement:* Barack Obama is black.
  Opposite: Barack Obama is white.
  Negation: Barack Obama is not black. In fact, his father was black and his mother white, so he is arguably as much black as he is white, or perhaps neither.

Thinking about opposites instead of negations is an extreme and very black-and-white way of looking at things, figuratively or otherwise. In the case of Barack Obama, most people would feel strange calling him white, and he is generally called black although he is in some sense equally both. So why does it seem to make more sense to call him black than white? Does it in fact make any sense? This is part of a question about gray areas.

GRAY AREAS

People are often not very good at dealing with gray areas. There are many arguments in real life that turn into arguments about opposite extremes. If two people go to a concert together one might enthuse "That was great!" and the other refutes with "How can you think that? I thought it was terrible." Political decisions usually result in some people saying it's a great decision and everyone else saying it's terrible.

People argue about whether a certain leader was good or bad, with people on one side mentioning all the good things they did, and people on the other mentioning all the bad things they did. In reality, most people do some good things and some bad things. In fact, most things are themselves partly good and partly bad. A more logical negation would be if someone was arguing that a leader did some good, and the other side argued that the leader did not do some good, i.e., did no good at all (which is very extreme). Or if one side argued that a leader was utterly evil, and someone else pointed out that they were evil overall but also did some good at some point. Unfortunately, if you don't condemn every single thing a person did then it can sound to some (less logical) people that you are supporting them. This is the trouble with black and white thinking.

People are not very good at dealing with gray areas, and in fact nor is logic. We'll come back to this later (in Chapter 12) but for now it's important to note that the gray area should be included *somewhere*, otherwise we're just ignoring part of reality.

In a formal debate situation it is clearly set out where the gray area is: it is included in the opposition. So with the strawberries and raspberries debate, the motion is that strawberries are "definitely" better than raspberries. All the gray areas are in the opposition: the two types of berry could be roughly equal, they could be sometimes better and sometimes worse, and so on.

If we're thinking about the concept of "good", then the gray (mediocre) is included in "not good", so it gets lumped in with "bad". If we're thinking about whether or not the EU is terrible, then the gray (so-so) is included with "not terrible", so it gets lumped in with "fantastic".

If we're looking at the (already flawed) notion of race and thinking about white people, then all shades of gray are included with black in "not white". This idea was enacted in some parts of the US in the twentieth century when anyone with just one drop of "black blood" was considered to be black. At other times the arbitrary cut-off point was picked to be one eighth or one quarter ancestry.

If we only talk about black people and white people then we have either picked an arbitrary cut-off point as above, or actually erased everyone else from our discussion, including mixed-race people, Asians, Native Americans, and anyone else who is neither black nor white.

In logic this is called the Law of the Excluded Middle. It says that "true" and "not true" are the only two options we are going to deal with at the moment. So all types of "not true" have to be included. It also means that if something is not not true then it has to be true. It doesn't mean we've excluded the middle in the sense of throwing it away or ignoring it, it just means we've included it with one side or the other so that there is effectively no middle any more.

Here is a picture of a sliding scale from white to black:

Where should we draw the line between white and black? One strictly logical approach is to consider black and not black:

not black   black

But another way that is just as logical is to consider white and not white, in which case the line is towards the opposite end:

white   not white

There is much that is unsatisfactory about these approaches as in both cases the line has been pushed to one extreme. When considering racial issues it can be productive to talk about white people and non-white people as the privilege of white people does not seem to extend to mixed-race people unless they can "pass" as white. However, on the other hand, calling everyone who is non-white some sort of "other" can be a symptom of white supremacy and the reluctance of white people to let others into their part of society.

Pushing the lines to the extremes is at least more logically sound than considering only the extremes and ignoring the whole gray area – after all, if we act as if the middle doesn't exist then what we are saying is simply untrue.

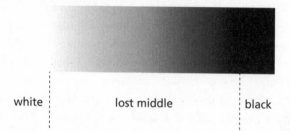

white   lost middle   black

Another less extreme but less logical approach that happens a lot in real life is that we place the line abitrarily in the middle somewhere:

white(ish)   black(ish)

Yet another alternative is to designate an area in the middle and call it something like "gray":

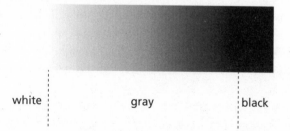

white   gray   black

Of course, this still means we have to pick a place to put the lines between white and gray, and between gray and black. This is to some degree what happens with the terminology of "heterosexual", "homosexual" and "bisexual". One extreme consists of people who are sexually attracted exclusively to those of the opposite sex, and the other extreme those who are sexually attracted exclusively to those of the same sex. The middle is then those who are sexually attracted to both. However, where do we draw the lines? If someone is attracted to those of the opposite sex but also to one person of the same sex, is that enough to call them bisexual? What if someone is attracted to those of the same sex but also to one person of the opposite sex – are they in that case not gay? The simple answer, to me, is that everyone is entitled to call themselves what they choose, but, as with considerations of race, there is an under-current of power imbalance pulling these considerations in one direction: the scale is not symmetrical. After a long history of oppression, black people and gay people have a lot more at stake in these classification systems, whether it's the need to assert one's identity or hide it, the need to protect one's community or the need to be allowed to integrate into the power-holding community as everyone deserves. While logic can helpfully simplify situations by ignoring irrelevant details, we should be careful not to oversimplify by ignoring important aspects of context.

In Chapter 12 we will come back to this issue and see that although the approach of placing lines somewhere in the middle is less extreme, and at least doesn't ignore the entire gray area, it causes other contradictions instead. We will discuss more nuanced ways of dealing with gray areas with the aim of simultaneously avoiding extremes, ignorance and contradic-tions. Where gray areas are concerned this is easier said than done.

Absorbing the gray area into one side or the other is a simplification, but at least not incorrect or contradictory. Whereas denying its existence altogether is where black and white thinking usually goes wrong.

VENN DIAGRAMS

To think about how to consider negation, consider white people and non-white people. We could have a Venn diagram of people, with a region for white people like this:

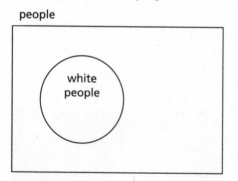

The outside part is then where we have all the people who are not white, shaded in here:

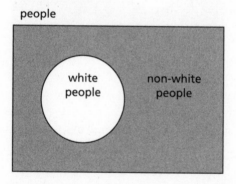

The non-white part then includes black people, Asian people, Latin people, Native American people, and anyone else who isn't white.

In general we could think about a statement *"A is true"* and draw a Venn diagram like this:

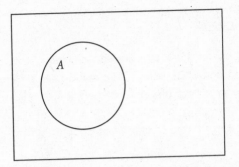

In that case the region where "A is not true" is the outside of the A circle.

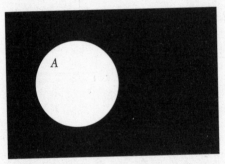

Here we come up against the problem of gray areas again, and the fact that we're using the law of the excluded middle: the negation takes up all the space outside the circle, and there is nothing in between that space and the circle itself. In the language of sets and Venn diagrams this is called the complement – the part that fits together with A perfectly. If we had a gray area it might look like this:

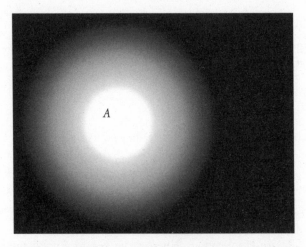

Now, if the concepts we are thinking about are "white" and "not white", the dividing circle is in the middle:

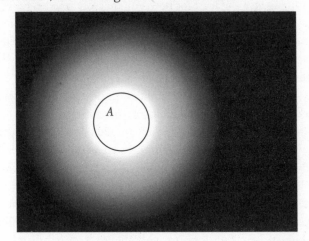

Whereas if we are thinking about "black" and "not black" the dividing circle is at the outside, just where the deep black starts.[1]

---

[1] You might be able to see an optical illusion at work in the first diagram (without the inner black circle). There seems to be a white glowing circle outside the A, with the central part looking more gray. This is an illusion: the whole centre really is white, but somehow our eyes interpret the outer ring as

We will see Venn diagrams with more sets in the next chapter when we think about how logical statements are connnected.

TRUTH VALUES

Mathematicians might seem like they're trying to make things more complicated all the time, but really they're trying to make things simpler so that we can try to understand them better. There's an important difference between simple and simplistic though, and I think it's to do with illumination. If you make things simplistic you are probably ignoring crucial details that actually are illuminating. Whereas the key to a good simplification is to keep some pertinent illuminating details and forget everything else – at least for now. Another key is to be aware of what you're forgetting all the time, rather like deliberately leaving your umbrella at home if the weather forecast is fine, rather than accidentally forgetting it and also forgetting to look at the forecast. If you're aware of what you're forgetting then you can also be aware of the limitations of what you're doing, and the situations you shouldn't get yourself into.

In a way the law of the excluded middle is a simplification. We will later see that this means there are certain situations we really can't deal with without modifying our logic. Another way of viewing this law is that we are viewing truth as binary, just "yes" or "no". Mathematicians push this further and give truth a value: 0 if something is false, and 1 if it is true. You might think mathematicians just love turning things into numbers, but remember: math isn't just about numbers, it's about many other things too. However, numbers are so familiar and easy to

---

more white, probably because of its proximity to the gray. I feel like there might be a metaphorical interpretation here, that we also think of there being a demarcation between white and non-white in our heads even when it's a sliding scale, especially if we're too used to being in the middle of the white. We'll come back to this question as it pertains to racial prejudice later.

reason with that if we can turn a situation into some numbers it can be very helpful.

The law of the excluded middle says that there is no truth value in between 0 and 1. You might cry foul at this point and say doesn't this mean we have thrown away the middle? There are, after all, a lot of decimal numbers between 0 and 1. These might try to encapsulate partial truth. Perhaps if something has truth value 0.5 that means it is half true? There is a form of logic that takes this approach, called "fuzzy logic", and we'll come back to it in Chapter 12 when we discuss more ways to deal with gray areas.

Using 0 and 1 as the only possible truth values is like allowing only the answers "Yes" and "No" in court, a favorite tool of lawyers when they're pressuring someone to say something that casts them in a bad light (at least, this is what fictional lawyers do). In logic, and in court, things are simply true or not true, 1 or 0. This can sound draconian, which is why it's important to remember that "not true" does include all possible shades of gray.

If a statement is true then its negation must be false. Also, if something is false then its negation must be true. We can sum this up in this little truth table. Here A is any statement, and "not A" is its negation:

A	not A
1	0
0	1

Truth tables, like Venn diagrams, are another useful way to encapsulate logic. They get a little bit further away from intuition, but sometimes when you move away from intuition two things happen. One is that you are better able to engage your logical brain. The other is that you actually develop new intuition, intuition about logic rather than anything else. If logical intuition sounds like a contradiction then I apologize. (And that's a promise rather than a logical implication.)

Finally I should note that there is one other possibility for truth values: it is possible that the truth value for something cannot be determined. We'll come back to this in Chapter 9 on paradoxes. Some paradoxes are caused by a statement that is so self-contradictory that it can't be given any truth value – true or false – without causing a contradiction. That means it can't have any truth value at all. It doesn't mean it's false, it means that it is unknowable.

Here are some statements whose truth value is currently unknown, but only as a result of the limitations of current human knowledge, not because of internal logical problems:

1. The universe is finite.

2. One day we will be able to cure all cancer.

3. A meteor caused the extinction of the dinosaurs.

NEGATING IMPLICATION

Now that we understand more about negation we can try applying it to a more complicated statement: a statement of implication.

We have seen the implication "If you are white then you have privilege". We have seen that the converse "If you have privilege then you are white" isn't true because you could be non-white but have other forms of privilege, for example by being male, rich, straight, cisgendered,[2] able-bodied, tall, thin, educated. Since it isn't true, we should be able to negate the statement.

It is tricky to negate implication. There isn't a simple way to negate it by just sticking a "not" into the sentence, although it is

---

[2] Cisgendered: your personal gender identity matches the sex you were assigned at birth.

tempting. We can try to say "If you have privilege then you are not white" which isn't true. If you have privilege you *might* be white, but you might not.

We might try to say something like "There are other types of privilege besides white privilege". This is a good step towards a logical negation, but we haven't talked about "There are ..." and what this means yet. (We will in Chapter 7.)

Until then, really the only logical way to negate "implies" is by saying "does not imply", as in "Having privilege does not imply that you are white". Or we can just tack "It is not true that ..." onto the beginning of the statement. This is a foolproof way of negating a statement, it just tends to result in an unnatural sounding sentence. "It is not true that if you have privilege then you are white".

In symbols we just sort of cross out the implication arrow and write

$$A \nRightarrow B.$$

Implications that are not true are often at the root of disagreements.

FAULTY IMPLICATIONS

Here is a faulty argument that some white people use to try and argue that white privilege doesn't exist:

> "Some black people are better off than me,
> therefore I don't have white privilege."

Now, it is true that some black people are better off than you, even if you are white; for example, unless you're very unusual Barack Obama and Oprah Winfrey are probably better off than you. However, this does not mean you don't have white privilege.

Here is how that faulty argument tries to proceed, in more detail:

1. Some black people are better off than me even though I am white.

2. If some black people are better off than you then you don't have white privilege.

3. Therefore I do not have white privilege.

The process of concluding something from a logical implication is called inference. This rule of inference is very fundamental to the use of logic, and has a fancy name: *modus ponens*, which is Latin for "way of affirming". It is essentially the only way we can progress from one known truth to another one. (In Chapter 9 we will look at Carroll's paradox which explores the impossible state we would get into if we were not allowed to use this rule of inference.) Modus ponens says that if we know *"A implies B"* then we can *infer B* from A as follows:

1. *A* is true.

2. *A* implies *B*.

3. Therefore *B* is true.

In the above example the conclusion is "I do not have white privilege." There are now two ways the conclusion could be false: either (1) is false, which would mean that you are better off than *all* black people (or you are not white) or (2), the implication itself, is false. For our example some people think (2) is true, but this is a misunderstanding of what white privilege means. It is a straw man fallacy, that is, when an argument is replaced by one that is not the same but is much easier to knock down (a straw man) and is then duly knocked down. (We will come back to straw man arguments in Chapter 14 on equivalence.) White privilege doesn't mean that every white person is better off than every non-white person; it means that if any given non-white person were in exactly the same circumstances but white, they would be better placed in society and life.

The key is to remember that it is *both* statements *"A"* and *"A* implies *B*" that *together* allow us to infer statement *B*. So if statement *B* is not true it is either because *A* isn't true or because "*A* implies *B*" isn't true. The possibility of the implication being untrue is often overlooked. In the next chapter we will look more closely at how factors combine to produce results, and how some factors are often overlooked so that blame is unfairly focused on one particular person or circumstance.

CONTRAPOSITIVE

I believe that a great deal of logical power comes from flexibility, and this comes from being able to see things from different but equivalent points of view. Understanding how negation interacts with implication gives us a way of doing that. In the above argument, believing that

> "If some black people are better off than a white person
> then that person does not have white privilege."

is equivalent to believing this:

> "If you have white privilege
> then you are better off than all black people."

The second statement is logically equivalent to the first. This means that the second one follows from the first, but also the first one follows from the second, so they are logically interchangeable (but they might still give slightly different emphasis to us non-logical humans.) For example, suppose I tell you:

> If you travel abroad you must have a passport.

This is logically equivalent to saying:

> If you don't have a passport you can't travel abroad.

However, they are perhaps emotionally a bit different – the first version is more about what you need to do in order to travel whereas the second is more about what you can't do without a passport. These are two slightly different points in human terms although they're logically equivalent.

These pairs of equivalent implications are called "contra-positives". Formally, the contrapositive is a new statement that you can make, starting from an implication

$$A \implies B$$

The contrapositive statement is then

$$B \text{ is false} \implies A \text{ is false}$$

and is logically equivalent to the original statement. This means that whenever the original is true, the contrapositive is true. And whenever the original is false, the contrapositive is false. This is not to be confused with the converse, which is

$$B \implies A$$

and is logically independent of the original. It is also not to be confused with what you get if you just negate A and B individually:

$$A \text{ is false} \implies B \text{ is false}$$

which is the contrapositive of the converse, and so equivalent to the converse.

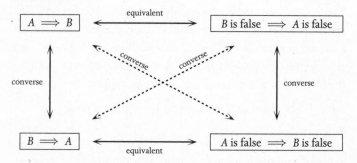

Negating A and B individually is the mistake people are making when they look at the statement

If you are a US citizen then you can legally live in the US.

and turn it into

If you are not a US citizen then you can't legally live in
the US.

Finally, the contrapositive is not to be confused with the negation:

$$A \not\Rightarrow B$$

which does not fit into the above picture anywhere. If an implication is false you can't deduce anything from anything else, except that the argument is broken.

EVIDENCE

A logical implication is much more powerful than a piece of evidence. A logical implication means that something is *definitely* true. A piece of evidence only contributes to the probability that something is true. This is an important difference. Evidence can't contribute to the logic of something being true; only a logical justification can do that.

For example, suppose you think all Chinese people are good at math. Perhaps every time you see a Chinese-looking math person (like me) you think this adds to your theory. You are thinking of the implication

Being of Chinese origin implies being good at math.

Every time you meet a Chinese-looking mathematician you store this up as evidence. One issue is that you might be succumbing to confirmation bias, where you only notice evidence that backs up your theory. However, another issue is simply that evidence doesn't contribute to logic. With our new power in dealing with contrapositive statements, we can perhaps see this more clearly. The contrapositive of the statement is

Not being good at math implies not being Chinese.

Working by the idea that "evidence contributes to logic", this means that every time you see a non-Chinese non-mathematician this should also add to your theory. For example, every time you see a Canadian goose, or a French macaron, this should add to your theory that all Chinese people are good at math. You might object that we're only thinking about people here, not animals or food. But that still means that if you meet an American singer or a British football player, that should add to your theory that all Chinese people are good at math.

Intuitively this might seem much more bizarre than adding to your theory every time you meet a Chinese mathematician, but logically speaking both cases make equally little sense.

NEGATION IN SCIENCE

While we're thinking about evidence, it is just as well to think about what evidence *does* do for us in scientific experiments, and how it interacts with negation. After all science, like mathematics and all fields, has a framework for showing that things are not true as well as for showing that they are true.

Scientific experiments usually proceed by starting with a hypothesis, that is, a statement that a scientist thinks might be true, but whose truth value is currently unknown. In school science experiments it usually has to be something rather simple that scientists have in reality known for a very long time. This always made me feel that science experiments at school were contrived. Because of my lack of motivation for doing the experiments, I was not very good at them. I wish they had been presented more as a way of exploring the scientific method and learning how to verify things that have been told to us as facts.

One experiment that I remember actually going well was the one about Hooke's law. The hypothesis is: the extension of a spring is proportional to the load that is hanging from it.

Scientists (or school pupils) will then look for evidence to support the hypothesis. In this case, an experiment would involve selecting various different springs and measuring them with various different loads hanging from them, and then analysing the data. The field of statistics deals with what sort of data you need to back up various types of hypotheses. If you find you have the right kind of data then you make a logical conclusion of this form:

There is sufficient evidence to suggest that this hypothesis is true, to within 95 percent certainty (for example).

If you find you do not have the right data then your logical conclusion should be the negation of this statement, which is:

There is not sufficient evidence to suggest that this hypothesis is true, to within 95 percent certainty.

It is important to note that this is logically different from concluding that the hypothesis is false, which would be the opposite rather than the negation. If you have insufficient evidence to support a hypothesis, it means that the truth value is still unknown. Maybe you need more data. Maybe you need a better experiment. In the example above what we really need is a refined hypothesis:

The extension of a spring is proportional to the load hanging from it, within a maximum load limit.

This is Hooke's law of springs. Once the truth of a hypothesis has been scientifically established it typically gets "promoted" to the status of a law. A scientific law is something that has been determined to be probably true, to within levels of certainty accepted by science. This is different from logical truth. However, logical truth interacts with scientific truth: we proceed logically from the scientific law by saying if the law is true then various things can be deduced from it logically.

Sometimes people point to the percentage certainty involved

with scientific experiments and declare that this shows it's "only a theory" and therefore we're entitled not to believe it. This is misunderstanding the scientific method, statistics and probability. If something is found to be true with 50 percent certainty then indeed the jury is out and it could go either way, and we'd probably be wise not to act based on either its truth or its falsehood, but to await further input. However, if something is found to be true with 95 percent certainty then it is very likely to be true, although there is still a small chance it is not true. Scientists pick their percentage boundaries based on the seriousness of the situation – again it's a question of false positives and false negatives. Would it be worse to say it's true when it isn't, or say it's false when it's true? For life-threatening situations like drug side effects, a higher level of certainty is used, but because absolute certainty is never possible outside of logic, there will never be 100 percent certainty. But if we only acted based on things we knew with 100 percent certainty, we would rarely do anything at all.

On the other hand if something is deemed to be true with 1 percent certainty then it is indeed quite certain to be false, but even then it could still be true. My excellent math teacher, Mr. Muddle, taught us that when you work as a professional statistician, if you do not have the right data to support your hypothesis the correct negation is "There is insufficient evidence to support this hypothesis and therefore we need more funding in order to pursue the matter further."

5

BLAME AND RESPONSIBILITY

HOW EVERYONE AND EVERYTHING IS LOGICALLY CONNECTED

ON 9 APRIL 2017 UNITED EXPRESS Flight 3411 was overbooked. The airline bumped a particular passenger off the flight, but he didn't go voluntarily and was dragged off by security officers, sustaining injuries along the way. There was an uproar and opinion was characteristically divided about whose fault it was. The main opposing viewpoints were:

1. It was United's fault for their unreasonable use of force.

2. It was the passenger's fault for refusing to leave his seat when asked.

But there were very many contributing factors. Is a contributing factor the same as "fault"? Let's face it, everything in life is caused by more than one factor. It's just humans who are prone to trying to point the finger of blame at one factor, often one person.

If a student doesn't do well in an exam, is it because they didn't work hard enough, or because they were not taught well enough? Probably it is both of those things to some extent: a really excellent teacher will inspire students to work hard, but this sounds like placing blame on the teacher; a really good student will work hard even if they have an uninspiring teacher, but this sounds like blaming the student for being taught badly. There's a cartoon that goes around periodically in which the "good old days" (whenever those were) are compared with today. In the panel for the good old days, a parent and child are in a teacher's office, and the teacher is scolding the student for their bad grades. In the panel for today, the image is the same except now it's the parent scolding the teacher for the student's bad grades. There is, alas, some truth to this. The question of blame is wrapped up with the question of responsibility, and the counterargument is often that if we don't blame any individuals, does that mean that nobody should ever take responsibility for anything?

Another more universal case is when relationships break down. Sometimes it's mutual and both partners agree that it wasn't really anyone's fault, but alas this happens far too infrequently. Usually someone – or indeed both people – get very hurt, and each blames the other person. But in many cases (except in cases of abuse) there are contributing factors from both sides, and the key to understanding the breakdown is to understand the way in which the people were relating to one another, and the way their individual contributions were woven together to lead to the collapse.

In all these cases we do better to understand all the factors and how they are connected, and this is what we address in this chapter.

INTERCONNECTEDNESS

Going back to the example of the failing student, we can try to find the logic of the situation. The point is that the student's contribution and the teacher's contribution *combine* to cause the outcome:

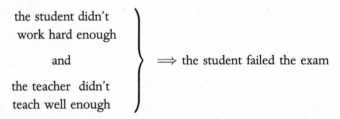

$$\left.\begin{array}{c} \text{the student didn't} \\ \text{work hard enough} \\ \\ \text{and} \\ \\ \text{the teacher didn't} \\ \text{teach well enough} \end{array}\right\} \implies \text{the student failed the exam}$$

Someone might argue "Well, if the student had worked harder, they'd have passed, so it's the student's fault." Someone else might argue "Well, the student did the best they could but they were so badly taught they didn't have a chance, so it's the teacher's fault". A former Oxford student recently sued Oxford University for lost income on the grounds that he was taught so badly it was Oxford's fault he didn't get a first-class degree, and that this caused him to lose income in his years since graduation. It's unadvisable to comment on such cases without all the information, but I would hope that there are better remedies for poor teaching than a lawsuit years later.

The crucial point is that when two factors combine to cause a result, either one of them being different could cause the result to be different. But that doesn't mean that one of them is individually to blame for the result: it is the combination of the two that caused the result. The logic of this situation is the logic of connectives.

Logical connectives are the way that logical statements are connected to form bigger, more complex statements. It is a general principle in mathematics that a good way to understand something complex is to break it down into simple constituent parts. Then you just have to understand simple building blocks together with the ways of sticking them together. Logical connectives are the way of sticking together simple logical statements into complex wholes.

For example, "The student didn't work hard enough and the teacher didn't teach well enough." The connecting word here is "and". How could the student have passed? Perhaps if "The student worked harder or the teacher taught better." The connecting word here is "or". These two words are the two basic connectives in logic.

"And" and "or" are innocuous little words, and yet they cause logical mistakes all over the place, especially when combined with implication and negation. In math they are called connectives because they connect up different statements to form new ones, like connecting pieces in a building set that join the rods together. Mathematicians love building complicated things from smaller things. The idea is that you can then understand really huge and complicated things, by understanding the small parts and the steps for putting them together. This is in general a good way of understanding a complex situation: break it down into small parts and carefully understand how to stick them together.

The general situation is that given two statements A and B, we get two new statements by using "and" and "or":

1. A and B are both true.

2. A or B is true.

As usual, some issues arise to do with how we use these words in everyday life.

In real life it might be that the word "and" doesn't explicitly appear in a sentence, but we can turn the sentence into an equivalent one where the connective has been made explicit. For example, if someone is a white man, they are white *and* also a man. Similarly "This is a racial slur" means "This is a statement about race and it is a slur". The long version sounds pedantic in normal language but clarifies the logic, so in the context of logic could count as precision rather than pedantry. Occasionally clarification is needed in normal language as well. I once said "He's a black cab driver" in the US and got some very funny looks because people thought I was saying "He's black and a cab driver", when really I meant "He's a driver of black cabs". (Black cabs are not really a concept in the US and so the interpretation with "and" was more obvious.) The first time I bought a packet of Yorkshire "Chardonnay wine vinegar crisps" I was amused to read the ingredients and discover it was Chardonnay wine *and* vinegar, not vinegar made from Chardonnay wine.

The mathematical concept of "or" is a bit more tricky than "and" because it's not quite the same as the way we use it in everyday life. If you ask someone "Would you like tea or coffee?" you might expect the answers tea, coffee or neither. If you ask a pedantic mathematician they are liable to answer yes or no. This is because in mathematics "or" is a logical connective that joins two statements *A* and *B* together to make a new statement "*A* or *B*". The new statement is true if either *A* or *B* is true, *or both*. This is different from normal usage in English where "or" often excludes the possibility of both – if I see "tea or coffee" on a fixed-price menu I expect to choose one, and expect that I won't be allowed to ask for both. This difference makes "or" logically ambiguous, and we humans tend to have to infer the correct meaning from the context. For example, you have to pay excess baggage fees on a plane if your luggage is too big or too heavy. It should be clear from the context that if it is both too heavy and too big you still have to pay extra. This makes it different from the "tea or coffee" type of situation.

Similarly you have higher social status if you are rich or male (or both). You are an LGBTQI person if you identify as lesbian, gay, bisexual, transgender, queer or intersex, or more than one (for example, transgender and lesbian).

Logic and mathematics require things to be unambiguous without us having to understand things according to context. So we have to make a difference between these two uses of the word "or". The one that excludes the possibility of both is called "exclusive or". When you're offered tea or coffee (but not both) that's an exclusive or. The one that includes the possibility of both is called "inclusive or", so when you have to pay more if your luggage is too heavy or too big (or both), it's an inclusive or. This distinction is usually clear from context in normal language, so it becomes pedantry rather than precision. However, in logic we have to make the distinction clear without guessing about context, so the distinction becomes precision rather than pedantry. So when a mathematician answers "yes" to the tea or coffee question, it means that yes, they want tea or coffee or both, and I would say they are being pedantic because not only are they not clarifying the situation, they are actively obfuscating it.

In math we tend to default to the inclusive or because it

makes things fit together better logically, as we're about to see. However, in normal life we tend to default to the exclusive or, although we sometimes emphasize it by using the word "either". As with implication and negation, we can picture "and" and "or" using Venn diagrams.

VENN DIAGRAMS

Suppose we are thinking about people who, on average, earn less than the general poppulation. Suppose black people suffer this disadvantage, as do women, but the people at the most grave disadvantage are those who are both black *and* women. We can depict this in a Venn diagram:

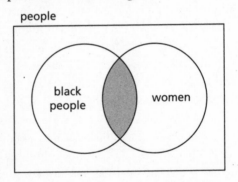

The region of overlap in the middle is where both things are true. In the language of sets and Venn diagrams this is called the intersection, and in this case it consists of black women.

If instead we consider those who are at some kind of income disadvantage (not necessarily the worst) then we need to include black people who are not women and women who are not black, as well as those are both. This gives us the region shown in the following diagram, which in the language of Venn diagrams is called the union of the two sets.

people

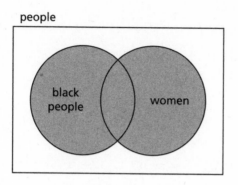

We will now see how "and" and "or" are related by negation.

NEGATION OF "AND" AND "OR"

We have already seen a few situations where two factors caused an outcome, and then people argue about which one is to blame. The thing is that the logic of *causing* an outcome is not the same as the logic of *preventing* an outcome. If two things are needed to cause an outcome, only one needs to change to prevent the outcome. In terms of logic, this is about negating a statement involving "and".

For example, if you're not a white man you could be a non-white man or you could be a white woman (or more generally a white non-male person); if we negate both the "white" and "man" parts we see that you could also be a non-white non-male person. We can see this in Venn diagrams as it is the region *outside* the intersection, as shaded here:

people

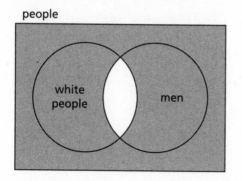

This consists of three regions:

1. white people who are not men,
2. men who are not white, and
3. people who are neither white nor men.

That is, it consists of people who are either not white *or* not male (or neither). In general:

> (A and B) is false means A is false or B is false (or both).

We can avoid having to say "or both" all the time by agreeing that we're talking about the *inclusive* or.

What about negating "A or B"? We can also go back to the question of income disadvantage: you are at an inherent disadvantage in Europe and the US if you are black or female. To escape this particular disadvantage, you must be not black *and* not female. (Of course, you might still have other disadvantages, such as being poor or sick.)

This is the very outside of the Venn diagram:

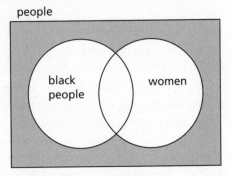

In general

> (A or B) is false means A is false and B is false.

To sum up:

- *Original statement:* You are black and female.
  Negation: You are not "black and female". So you could be black but not female, female but not black, or neither.

- *Original statement:* You are black or female.
  Negation: You are not "black or female". So you are not black and you are not female. In more natural language you are neither black nor female.

In both cases it has to be the inclusive or in order to make a satisfying relationship between "and" and "or", whether we're negating "and" and finding "or", or negating "or" to find "and". This is one of the reasons mathematicians prefer defaulting to this type of "or" – because it makes for this very neat relationship with "and".[1]

All of these Venn diagrams only really help us in very basic situations involving two (or perhaps at most three) sets. We are soon going to see that most situations have far more constituent parts like that, and drawing diagrams of the *flow* of logic is more illuminating.

The logic that we have at our fingertips so far is called propositional logic. It involves propositions (or statements), connectives and truth values. It is a bit simple, but still quite powerful in human terms for analysing issues of blame and responsibility, as we will now see.

BLAME

If factors *A* and *B* cause a situation, which is to blame? We now know that to negate an "and" statement we only have to negate one of the individual statements. This means that just one part could be to "blame" for the statement being false, although *both* of them have to contribute to the statement being true.

---

[1] To negate the exclusive or, we would have to get "both or neither". For example, you could find all the people who like tea or coffee but not both. The *other* people would be all the people who like both or neither. "Either . . . or . . ." is a fairly natural concept but "both or neither" is a bit strange.

If I break a glass I might say there were two contributing factors:

A: I dropped the glass.

B: The floor was hard.

It is the *combination* of those two things that caused the glass to break. Now, if I hadn't dropped the glass, it wouldn't have broken. But also, if the floor hadn't been hard, the glass wouldn't have broken. Negating just one of those factors negates the statement "*A* and *B*", but it doesn't mean that one of those factors is to blame by itself. It is the combination that is to blame.

In reality with the glass there are many other factors including the weakness of the glass and the action of gravity. We can string as many statements as we like together using the "and" connective, and the negation will always work the same way:

(*A* and *B* and *C* and *D*) is false

means

A is false or *B* is false or *C* is false or *D* is false.

That is, negating any one factor negates the whole statement. So for example, if you are not a straight, white, rich, cisgendered male person, it could be because you're not straight, not white, not rich, not cisgendered or not male. Losing any one of those privileges would mean you don't have the full complement of these types of privilege, but it doesn't mean that one of those privileges is more "to blame" than the others if you do have all that privilege.

This is a subtle point but a crucial one, I think, when we are considering who or what is to blame for something. For the student failing the exam we could also consider a lot of other factors: the exam was hard, the examiner was not lenient, the pass mark was set high, the student was sick that day. It is always easy to blame just one of those factors, on the grounds that changing one of them would have changed the result. But it is always the combination of those factors, joined by the logical connective "and" that actually causes the result.

I heard an interesting talk by software developer Jessica Kerr who summed this up as understanding the system rather than blaming the individual. So, instead of arguing about trying to attribute blame individually, it is more productive to understand how the system makes all those factors interact with each other to cause that outcome.

One of my favorite examples of this is from *An Inspector Calls* by J.B. Priestley. A woman has been found dead, and gradually more and more people are found to be implicated in her demise, in different ways, from personal to professional to incidental interactions. They all start arguing with each other about who is really to blame, when in fact it is an "and" situation. The mother and the father and the son and the daughter caused the situation between them, together with society and the world. Perhaps also it is an example of what Jessica Kerr said about understanding the system. Here there are two systems: the family and its interactions (for example, where Eric says "You are not the sort of father a chap could go to when he's in trouble."), and society and the way it treats poor women.

DIVORCE

You probably didn't expect to see a section on divorce in a book about math, but blame and responsibility is a central theme in many relationship breakdowns. If the breakdown is amicable, it is probably the absence of blaming tactics that is at root.

Consider a very simple but classic situation: someone has an affair. This doesn't automatically cause a relationship breakdown, so I would say "affair implies divorce" is not a fully logical statement. However, if one person has an affair and the other person doesn't forgive them, the relationship is doomed to break down. I'm going to call the protagonists Alex and Sam. We might have the following factors:

A: Alex had an affair.

B: Sam does not forgive Alex.

X: Alex and Sam split up.

Sam and Sam's friends probably blame Alex for having the affair. Perhaps Alex (and Alex's friends) blame Sam for not forgiving Alex. Maybe they think having an affair is bad but that nobody's perfect, and aren't you supposed to love someone for who they are, faults and all?

Really, of course it's *A* and *B* together that caused *X*. But maybe there are other factors as well, like:

C: they both refuse to go to couples therapy, or

D: they went but the couples therapist wasn't very good.

But we could also ask why Alex had the affair in the first place. Maybe it was because Alex is a liar and a cheat and a good-for-nothing, always out for a cheap thrill. Or maybe Alex was very unhappy because Sam was being neglectful. Why was Sam being neglectful? Maybe because Sam had become comfortable in the relationship and is generally a lazy and uncaring person, or maybe it was because Sam suffered a family tragedy and has been beset with grief. Or maybe it was because Alex was being cold. Why was Alex being cold? And so on.

Likewise, we can also ask why Sam does not forgive Alex. Maybe because Sam is a mean and ungenerous person who holds people to unreasonable standards. Or maybe it is because the way that Alex conducted the affair was so extremely hurtful that there is no chance of forgiveness. Or maybe it's because Sam has run out of goodwill because Alex has been treating Sam badly in other ways for some time. But why is that? And so on.

There is a general principle here in which one person does something to make *themselves* happy, but it causes the other person pain. Alex is unhappy so has an affair which brings some happiness, but it causes Sam pain. This is a type of zero-sum game, where one person can only gain if the other person loses. I think many toxic relationships boil down to this. Perhaps Alex feels stuck because if they do something to help themselves then Sam complains. Sam blames Alex for hurting them, but whose fault is it that Alex's happiness hurts Sam? The problem is the zero-sum relationship.

The upshot is that except in extreme cases of severe mistreatment and abuse, the situation is unlikely to come down to one factor, but rather, a complicated web of factors with interrelated implications and codependencies. We need to understand the system, and in this case the system is the relationship between the people.

THE EDUCATION SYSTEM

The education system is full of problems, in my opinion. There are problems to do with funding, expectations, objectives, standards, and so on. Problems with the education system is where I believe math phobia comes from. Many people develop math phobia at school because of the way it is taught. This sounds like I'm blaming the teachers, but is it really the teachers' fault? The teachers are under all sorts of pressure to meet arbitrary standards that are imposed on them. The standards are tested by exams under a time pressure, which means that teachers inevitably find themselves teaching "to the test" because they themselves are being judged by the test results. The other problem is that elementary school teachers are usually expected to teach everything, but if math wasn't something they particularly enjoyed at school, they might not be very comfortable teaching it in interesting and open-ended ways. This causes children to lose interest in it, often towards the end of elementary school, when the math has become too hard for general elementary teachers to be comfortable with, but specialist math teachers have not yet been invoked. This isn't the teachers' fault exactly, it's the system that places these demands on teachers.

Parents can also contribute to the math phobia of their children, by their own attitude to math, whether it's by pushing it excessively or by having their own phobia. But I believe those attitudes in turn come from the education system they experienced.

THE UNITED PASSENGER

With this new understanding of connected factors we can try analysing the abominable incident of the United Airlines passenger being dragged off the plane.

The simplest argument comes down to these two factors:

1. It was United's fault for their unreasonable use of force.
2. It was the passenger's fault for refusing to leave his seat when asked.

A slightly more nuanced argument declared it was the security officers who used the force, not United, so it was *their* fault. Some said it was actually United's fault for having the bad policy of asking passengers to leave at all. Simon Jenkins writing in the *Guardian* managed to blame everyone who travels. Somehow it's our fault for systematically putting up with bad treatment from airlines. Also it's the fault of passengers missing their flights that airlines have overbooking policies.

I think some of these arguments are more nuanced than others, but really, again, it's a case where all these are contributing factors. The situation was caused by:

1. The flight was unexpectedly over booked.
2. Some crew needed to get to Louisville to work on another flight, so the airline decided to remove people from the flight.

3. Nobody accepted the offer of money to get off the flight.
4. United decided not to offer more money.
5. United chose a particular passenger to be asked to leave the flight.
6. The passenger declined.
7. United staff called security officers.
8. The security officers used excessive force to remove the passenger.

To unpack this further we could wonder why the flight was so overbooked. Was it a holiday weekend? Why did the airline need to get those crew members to Louisville so urgently – had they not left enough leeway in their staffing model? Why did nobody accept the offer of money – was it not enough money or are people too greedy? Why didn't United offer more money – are they penny-pinching? Why did United choose that particular passenger? They say they choose based on airline elite status and "other factors". What were those other factors? Were they racially profiling? Is it significant that they picked an Asian man?

Why did the passenger decline? His stated reason is that he was a doctor and he needed to start his shift at the hospital. Do we then blame the hospital for giving him that shift? Do we blame United for picking a doctor of all people, whose arrival at work is arguably more critical than many other people's.

Why did staff call security officers – was the passenger being threatening or was United over-reacting? Likewise for why the security officers used excessive force. As for Simon Jenkins's argument, this seems to be an attempt to give reasons for the flight being overbooked and United having questionable policies.

We could draw a diagram of the logical interconnectedness and flow in this situation like this:

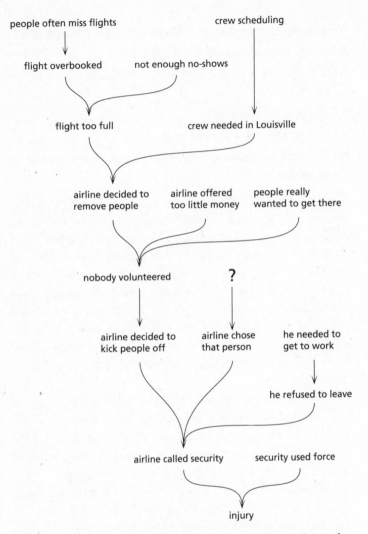

As you can see, asking careful "Why?" questions about the situation reveals a complex web of causation and shows how oversimplistic it is to try and blame just one factor. Simplistically blaming just one factor isn't the same as *simplifying* the situation to the most important factors though. I think identifying important factors is an aspect of powerful rational thinking, related to knowing when to stop asking or explaining "why" in any given situation. There is probably a whole PhD thesis that could be written analysing what happened on that flight and all the things that led up to it, including the entire way the airline industry works out how to send crew to the right places at the right times, the economics and statistics of overselling flights, and the psychology behind escalation to violence. However, if someone at a party said "What happened on that United flight? I didn't hear about it" it would be inappropriate to give that entire analysis, and indeed at the beginning of the chapter I didn't. We all know people who tell stories that are too long, or give explanations with way more detail than we wanted, so that our eyes start glazing over. Doing that is not exactly illogical, but it's not very *usefully* logical. On the other hand, oversimplifying and leaving out contributing factors is also not very useful. In online discussions, this is often preceded by the word "Fact:" or followed with "It really is that simple", when it rarely is that simple. Another phrase that can alert us to this sort of logical sloppiness is "End of story." Again, it rarely is the end of the story. For example, see if this sort of comment sounds familiar: "It's that guy's fault for not doing what he was told, end of story." Or "If you don't want to be injured being dragged off a flight, just get off the flight when you're told. It really is that simple." Or "Fact: if that guy had followed the crew's instructions he wouldn't have been hurt."

Many situations are much more complicated than this one. If we try to draw a diagram explaining the result of the 2016 US election we see a much more complex network of interconnecting factors:

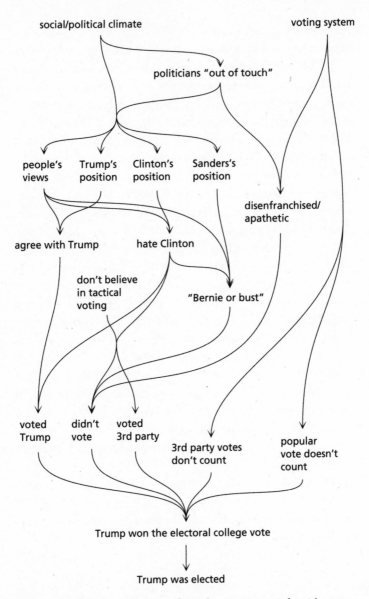

If I try and draw a diagram for why I gain weight, it's even more complicated because vicious cycles arise, as shown by the dotted arrows in the diagram opposite.

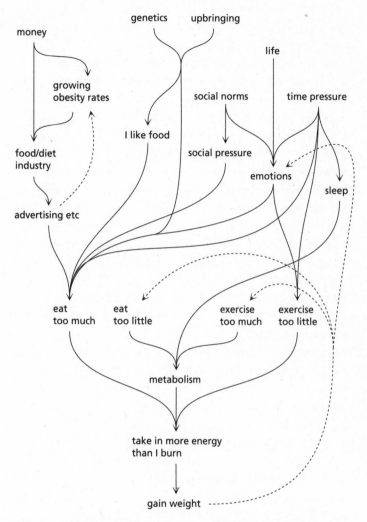

When people argue about who is to blame for something, it is usually an "and" situation: everyone involved collectively caused whatever it was to happen, linked by their particular, situational brand of "and". And usually whatever they did was in turn caused by something else, some other pressure in the system or society.

So who is to blame? It is possibly futile even trying to answer that question. A better question is: who is going to take responsibility for changing it? In a situation of jointly contributing factors, any one factor can change the end result.

Throughout the book I'm going to keep coming back to the point that the aim of intelligent rational humans shouldn't be simply to be logical, but rather, to be logical in a useful way. We could reply to the catchphrase "It really is that simple" by saying "No it isn't" and presenting a large diagram of interconnectedness as above. This risks merely complicating the situation, when an important part of understanding the complex world around us is simplifying it so that it is easier to understand. Yet simplifying by ignoring crucial factors is unfair. A better way of making something easier to understand is to become more intelligent. If we become better at understanding interconnectedness and systems, then instead of reducing a system to a single factor in order to understand it, we can be comfortable regarding the entire system as a single unit and understanding that.

Humans don't interact as cleanly as logical statements, but I still think it's illuminating to abstract them to logical statements to see that we're usually all responsible for everything together, because we live in a connected society. Unless you live in a cave. In which case you probably wouldn't be reading this book. It is tempting to point the finger of blame at one factor or person, especially if that exonerates ourselves, but I believe it is much more productive to understand the connections of the system. Outcomes are always caused by whole systems, but we can still as individuals take responsibility for change.

6

RELATIONSHIPS

IN THE PREVIOUS CHAPTER we saw how crucial it is to consider whole systems of interactions rather than people or events in isolation. The idea of considering things in relation to each other is one of the important basic principles of modern mathematics. This has not always been the emphasis, but relatively new research has brought it to the forefront since about the mid twentieth century. We see that looking at how things or people relate to one another is often the key to understanding a situation, more than looking at the intrinsic characteristics of those things or people. This is true at many different levels and scales, from how countries interact in the world, down to how people interact in a relationship.

VICIOUS CYCLES

One situation in which examining interactions is important is in vicious cycles, as we saw in the previous chapter in the diagram of why I gain weight. We can isolate a small part of that diagram to focus on the interaction between my emotions and gaining weight:

I feel bad   I overeat

I am definitely an emotional eater, prone to eating too much when I am stressed, upset or angry. But unfortunately overeating in turn causes me to be stressed, upset and angry. So I, like many people, get caught in a vicious cycle. Some people don't suffer from either arrow: emotions don't cause them to eat, but also, when they do overeat they don't feel bad about it. Suffering from one arrow is unfortunate but at least it doesn't cause the escalation that the two cause in conjunction. The cycle could be broken by breaking either of those arrows, which could be classified as "feelings" and "action":

I have tried breaking both of these arrows. When I am upset, I try to do something other than eat. Also, if I do find myself overeating I try not to feel bad about it. I find the latter easier to do than the former, especially when I'm stressed about some work I am trying to finish: I can stop myself from eating but then I won't get the work done, and I am far too responsible about meeting deadlines to accept that outcome.

The vicious cycle might have more arrows in it, as in a typical type of domestic meltdown between two people with slightly different needs. Perhaps Sam needs to feel loved where Alex needs to feel respected. While feeling respected Alex is able to be very loving, and while feeling loved Sam is able to show Alex plenty of respect. But if it starts going wrong it can escalate very fast, in the vicious cycle depicted below.

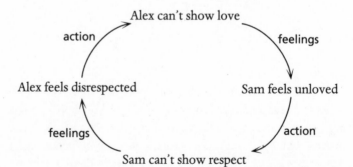

If they want to break the vicious cycle, one way is to look at which of these arrows is the easiest to break. As above, the arrows come in two types: feelings and action.

They might decide that feelings are hard to control, but actions can be changed. In that case the key is for Alex to show love even when feeling disrespected, or for Alex to show respect even when feeling unloved. Of course, there can still be an argument about who should break the cycle first. This can lead to very aggrieved people feeling that they are less at fault and that therefore the onus should not be on them to break the cycle. The best answer is that both people should make efforts to break the cycle. Perhaps an even better answer is one I read in *Love and Respect* by Dr Emerson Eggerichs, saying that whoever is more mature should break the cycle.

A more widely inflammatory example of the same principle involves police brutality against black people in the US. One highly simplified (but possibly illuminating) way of summing up that vicious cycle is this:

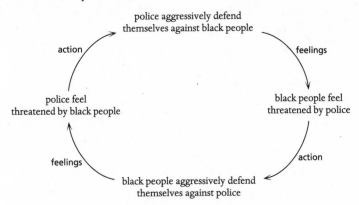

This is a very different situation but the question is analogous: if we want to break this vicious cycle, which arrow should be broken? Is it the case that actions are easier to change than feelings here too? Some people argue that black people should simply "do what they're told" by the police. But tragically, there are well-documented examples of black people being shot by police even when they were doing what they were told.

Other people argue that police should be trained to de-escalate all violent situations rather than responding with aggression themselves. One point of view is that whoever has the position of power should take responsibility for breaking this cycle. Some successful programs focus on changing both of the "feelings" arrows by fostering better relations between police and communities.

In all cases I think it's important to understand that this is a cycle, and claiming there is one root cause is an oversimplifi-cation, unless we acknowledge that the cycle itself is the root cause.

CATEGORY THEORY

Category theory is a field of modern mathematics that brings relationships between things to the forefront. In this approach, the framework for thinking starts with deciding what objects and relationships we're going to focus on. For example, when thinking about employees in a company, we could think about age, or we could think about years of service, or we could think about their position in the hierarchy of the company. Each of these would produce a different way of thinking about the interactions, and we could then think about what insight we get from these different lenses. We might discover, for example, that animosity arises when someone is higher up in the hierarchy but lower in age.

In category theory mathematicians discover that highlighting relationships between things is often much more illuminating than just thinking of things in isolation. Let's return for a moment to numbers. If we write down the factors of 30 we have these:

$$1, 2, 3, 5, 6, 10, 15, 30$$

Factors may elicit a shudder from those who only vaguely remember them from boring math lessons years ago. The truth is, I too think that a list of numbers in a line is boring. We live in a merely three-dimensional world, so we are constrained to write on two-dimensional pieces of paper, in one-dimensional straight lines. We often force our thoughts into one dimension when really they have natural geometry in higher dimensions. I like to say that this is why I don't tidy the papers on my desk – they have natural geometry in three-dimensional space where they are, positioned according to their relationships with each other by subject matter, importance, chronology, and so on. At least, that's my excuse.

We can find some natural geometry in our factors of 30 by thinking about which ones are also factors of each other. We can draw something like a family tree for them. As in a family tree, we won't draw lines between two "generations". We get this picture:

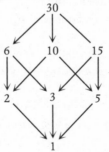

We now see that this has the structure of a cube – a more interesting structure than just some numbers listed in a straight line. We can then think about the hierarchy of these numbers in the picture. At the bottom we have 1, then we have the three next smallest factors, then the three next smallest, and the biggest one at the top. However, this hasn't happened because of the size of the numbers: it has happened because of which numbers are prime.

At the second level we have the factors 2, 3 and 5 because

nothing goes into them except 1. That is, they are prime numbers. (Recall that prime numbers are those numbers other than one that are only divisible by 1 and themselves.) The next level up has numbers that are each a product of two prime numbers:

$$6 = 2 \times 3$$
$$10 = 2 \times 5$$
$$15 = 3 \times 5$$

Finally at the top we have 30 which is a product of three prime numbers:

$$30 = 2 \times 3 \times 5$$

In fact, this structure arises precisely because 30 is a product of three different prime numbers. We could pick another number that is a product of three different prime numbers and do something similar, and see more clearly that the hierarchy is to do with prime factors, and not with size. For example:

$$42 = 2 \times 3 \times 7$$

has the following factors, in order of size:

$$1, 2, 3, 6, 7, 14, 21, 42.$$

If we arrange these in a picture as above it looks like this:

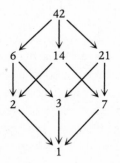

We now see that the next row of numbers up from 1 is *not* the next three smallest numbers: we have 2, 3 and 7 there as they are the three prime factors. The number 6 is smaller than 7, but has more prime factors as $6 = 2 \times 3$, so it's on the level above.

So we see that the hierarchy according to this diagram does not match the hierarchy of sheer size of numbers. If we also represented the size hierarchy, we would have to skew the diagram to look like this, a cuboid rather than a cube:

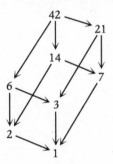

This might all seem like just messing around with numbers, but if we take it one step more abstract it suddenly becomes rather widely applicable. What we have seen is that the hierarchy comes from the prime factors of all the numbers, and this becomes clearer if we write down each number as a set of prime factors, instead of the number itself. For 30, we get this:

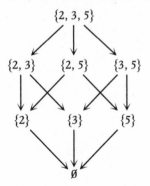

If we multiply the numbers in each position together, we get back to the actual diagram of factors from before. Here $\emptyset$ is used to represent the prime factors of 1, because 1 does not have any prime factors.[1]

For 42 we have this:

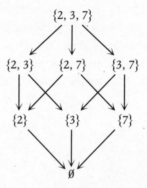

Comparing those diagramss we can see that every 5 in the first one just became a 7 in the second one. In fact, the arrows now represent the process of omitting an element from the set, so there is an arrow from $\{2, 3\}$ to $\{2\}$, for omitting the number 3, and one to $\{2\}$ for omitting the number 2, and so on.

Now, it doesn't matter exactly what these numbers were – the diagram would work for sets of *any* three objects a, b, c. So, more abstractly, we get this diagram of relationships between sets:

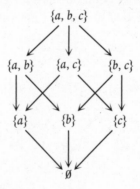

---

[1] The symbol $\emptyset$ is a standard one for the empty set. There are technical reasons that 1 is considered to be the "product of no prime factors".

We can now examine this diagram for *any* three things, and this is where the framework becomes very widely applicable.

PRIVILEGE

Consider three types of privilege: rich, white and male. Then, following the previous diagram of subsets, we get this diagram of relationships:

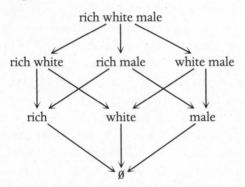

We can now add back in the full descriptions, for emphasis: contrasting male with non-male, white with non-white, and (for concision) rich with poor. We get the following picture:

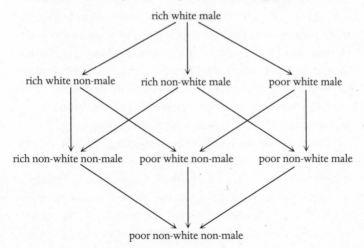

The first thing to observe is that this is a hierarchy where the layers show the number of types of privilege, rather than the actual amount of privilege. Thus of the types of privilege in question, the people in the top row have all three, the people in the next row down have two, the people in the next row down have just one, and those at the bottom have none.

Another thing to note is that the arrows show a direct loss of a type of privilege, with each direction representing one type of privilege. So the vertical arrows indicate the loss of privilege moving from white people to non-white people *who have the same other attributes*. This is an important aspect of privilege: it doesn't mean that all white people are more privileged than all non-white people, as can be seen from the fact that rich non-white males are, in this diagram, higher up than poor white non-males. Indeed, as we discussed in Chapter 4, some people point to the existence of super-rich black sports stars in the US and claim that this shows that "white privilege doesn't exist". However, the notion of white privilege doesn't mean that anyone is claiming that those super-rich sports stars are worse off than poor white homeless women. It just means that if those sports stars had all the same success but were also white, we'd expect them to have even higher status in society. This is also relevant to our previous example of police brutality against black people. Some people argue that the tragic deaths occur because the victim was doing something wrong, not because they were black. But if white people doing the same thing (or worse things) are not shot, then it is evidence of white privilege at play: where outcomes are improved if all circumstances are the same except for the person being white instead of black.

There is more we can learn from this diagram. Just as in the factors of 42, there is a tension between the hierarchy in this diagram and the hierarchy of absolute amounts of privilege. If we examine the second row in the diagram we might notice that the three groups of people on that second row do not have equal status in society. Many people would argue that rich white women have higher status than rich black men, for example, and that rich black men in turn are better off in society than poor white men (not just in terms of wealth). It turns out that money goes a long way towards mitigating other problems. Also, although women still suffer disadvantages compared with men, white women were granted some status in society long before black people, certainly in the US, and that history has left an ongoing legacy on society.

Anyway it is not our purpose to examine the historical causes of inequality, but to look at the logic of the situation. We could skew the diagram to show that the three groups in the second row are not level, and likewise the third row. But we can take this further. Instead of just comparing absolute privilege within each row, we can even compare between the rows. Arguably, wealth can outweigh being non-white and being non-male. If we think about rich black women, for example, we might well conclude that they are probably better off than poor white men. Consider Oprah Winfrey or Michelle Obama, for some extreme examples, and compare them with poor white unemployed men. So, in fact, the diagram is even more skewed, like this perhaps:

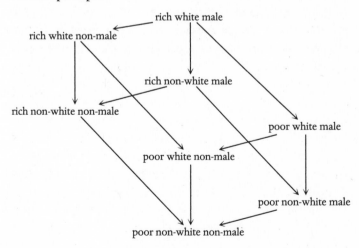

This diagram shows the tension between two different ways of measuring relative privilege: the number of types of privilege, which is shown by the arrows, and the absolute amount of overall privilege, which is shown by the height on the page. The disagreement between these two points of view causes animosity. In particular, this provides a logic-based account of why some poor white men are so angry in the current socio-political climate – because they are considered to be privileged from the point of view of number of types of privilege (white and male), but they are in reality less advantaged than many people who count as having fewer types of privilege than them. Understanding the root of this complaint is more productive than simply being angry with them in return. The same is true for other underprivileged groups within white men: any who are not rich, cis, straight, able-bodied, and any other types of privilege we can think of.

Diagrams of inter-relationships also help us to focus on the context we're thinking about, and in doing so, can help us switch context and be clear of how switching context affects inter-relationships. For example, instead of considering "rich, white, male" as three attributes, we could change our context to women, and consider "rich, white, cis" as three types of privilege. Leaving aside questions of absolute privilege for the moment, this produces an analogous cube of privilege as shown:

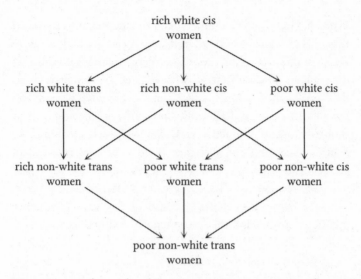

Here we see that rich white cis women now occupy the top position, analogous to rich white men in the previous cube of privilege. This helps us understand why there is a lot of anger towards rich white cis women, among women activitists who feel excluded by mainstream feminism. Rich white cis women may perceive themselves as being underprivileged, especially if they mostly inhabit a world that consists entirely of rich white people. At the same time they are viewed as maximally privileged in the context of women, or feminism.

These questions can become an argument about whether women standing together in unity amounts to erasing the experiences of the most disadvantaged women. From the other point of view, it can become an argument about whether taking into account the greater disadvantages of some women amounts to women fighting each other rather than the true opposition.

Category theory tells us it is always important to be clear what context we're thinking about. Everyone is privileged relative to some contexts and underprivileged relative to some others. Animosity tends to occur when someone is prone to thinking of themselves in a context that makes them underprivileged (a "victim") while others tend to view them in a context that makes them overprivileged. We need to find ways to attribute one type of privilege to someone without invalidating their feeling of disadvantage with respect to another type, as that invalidation causes anger, animosity, divisiveness and resistance to change. Rather, this possibility of flipping context should be put to productive ends. We will discuss this in more depth in Chapter 13 on analogies. If we all become more adept at seeing things from both a privileged and a non-privileged point of view, we will achieve greater understanding of disadvantaged people's struggles but also of the actions, whether malicious or ignorant, that cause bigotry and oppression.

7

HOW TO BE RIGHT

HOW WE MAKE OURSELVES MORE ACCURATE BY LIMITING OUR SCOPE

IN THE LOVELY STOP MOTION film *Chicken Run*, the smooth-talking American rooster Rocky does a deal with the sly salesmen rats saying he will pay them "all the eggs he lays that month". The rats are sly but not very knowledgeable about chickens, so they don't realize that roosters don't lay eggs, and that the total number of eggs Rocky lays that month is therefore going to be zero. Rocky is being entirely truthful in telling them that he will give them "all" those eggs, because "all" happens to be zero. Ginger, the heroic leader of the hens, is outraged and thinks Rocky has been dishonest. Rocky has certainly misled the rats but has not been strictly dishonest in terms of logic; he has just been dishonest in terms of suggestion or emotion.

A reverse emotional–logical situation occurs when someone has an emotional outburst and wails "Men are sexist pigs!". Do they mean all men are sexist pigs? That seems rather extreme. Do they mean most men are sexist pigs? Still a bit melodramatic. Perhaps we can agree that *some* men are sexist pigs? This is now a true statement but it's become rather tame – no wonder the wailer preferred the dramatic outburst.

Probably, in practice, they mean "I have encountered enough men who were sexist pigs today that I am fed up with it." This takes longer to say but is more precise. It also sounds a bit pedantic, but perhaps it is actually illuminating: it expresses the fact that actually what's going on is your emotional response of being fed up. However, to express that fed-up-ness it may be tempting to say "All men are sexist pigs!" but this might provoke someone to argue that not every man is a sexist pig, say, perhaps, Justin Trudeau. In this case you've expressed true emotions but with inaccurate logic, and in doing so you've tempted certain types of people to argue with your logic instead of soothe your emotions.

A toxic form of this kind of outburst is when someone accuses their partner of something like "You *never* do the washing up!" or "You *always* leave a mess in the kitchen!" The logical refutations of these statements are rather simple to prove:

- Statement: You never do the washing up!
  Negation: I did the washing up one time.

- Statement: You always leave a mess in the kitchen.
  Negation: There was one time I did not leave a mess in the kitchen.

Of course, the sweeping statement isn't meant literally. More precisely, it probably means something like "I feel like you do vastly less than your fair share of the washing up, so little that it feels negligible, so I'm very frustrated and feel overworked and underappreciated." Or "I feel like you leave a mess in the kitchen so often that cleaning it up becomes a big burden on me and I am tired of it."

It would probably be not just more accurate but also more productive to say those things instead of the exasperated sweeping statements.

SWEEPING STATEMENTS

People are prone to making sweeping statements. See, I just did it myself. What did I mean? Did I mean that *all* people are prone to making sweeping statements? Do I mean *some* people are? This is certainly true, but hardly a very emphatic statement. Do I mean that *most* people are? I think everyone I know is, but I've only met a tiny proportion of all people, so really I should just say "Everyone I know is prone to making sweeping state-ments". I have now refined my statement and made it less ambiguous, hence more defensible using logic. The way I did it was by refining its scope. Refining your scope means being precise about what world of objects you are focusing on.

"Mozart is more boring than Brahms" is a sweeping state-ment that many people would disagree with. Whereas "In my opinion Mozart is more boring than Brahms" is now just a statement about my taste, so nobody can logically disagree with it. It would be even more precise to say "In my opinion almost all Mozart is more boring than almost all Brahms." I can probably think of one example of a piece of Mozart that I think is less boring than one example of a piece of Brahms: for example, I am not a huge fan of Brahms's second symphony but I like Mozart's *Masonic Funeral Music*. Whereas I can make a sweeping statement "The macarons in Paris are much better than the macarons in Chicago" by which I mean "In my experience every macaron I have bought in Paris was better than every macaron I have bought in Chicago." Harsh, but true. Also impossible for someone other than me to refute logically.

"Almost all" is a form of qualifier that has a whole related family, like "most", "some". If you qualify a statement with one of those, together with perhaps "In my experience", you can almost never be wrong. (See?) "Perhaps" also works, along with "probably", "maybe" and "might". Carefully qualifed state-ments with these words are very precise in their correctness, but do not make good headlines, so unfortunately the media tends to overstate everything. "New research shows that sugar causes cancer!" screams a headline, but if you look at the research what it shows is that there is evidence to suggest that there might be some kind of a connection between sugar consumption and cancer. We could also add "seems", as in "It seems like there might be some kind of a connection between sugar and cancer."

Looking for the truth in someone's statement can be much

more productive than pedantically demonstrating that they are wrong. I think this is an instance of the principle of charity, where you try to think the best of everyone all the time. Finding truth in someone's sweeping statement by applying the right qualifiers can lead to greater understanding of what people are trying to say and where disagreements are coming from.

For example, in a typical argument about homeopathy, someone says that there is no evidence that homeopathy works, and someone else insists that homeopathic remedies make them feel better. The probable truth behind these statements is that there is no scientific research showing that homeopathy works any better than placebo, and the person insisting that homeopathy works for them is probably benefiting from the placebo effect. The placebo effect certainly has been demonstrated to work better than nothing. So the anti-homeopathy person is comparing homeopathy with placebo, whereas the pro-homeopathy person is comparing homeopathy with nothing. "No better than placebo" does not contradict "better than nothing", and so there isn't a logical disagreement here, probably just an emotional one about whether it's worth paying for something that is "just" placebo.

THE CERTAINTY OF LOGIC

Basic logic doesn't handle these nuances much better than we do in emotional outbursts. We've discussed gray areas and the way that basic logic forces us to push all the gray to one side or the other. When it comes to qualifying statements, there are two logically unambiguous ways to do it:

1. The statement is true of everything in your world. Perhaps "all mathematicians are awkward".

2. The statement is true of at least one thing in your world. Perhaps "there is at least one mathematician who is friendly". (I hope I count as this.)

Compare these two statements:

> Everyone in the US is obese.
> Someone in the US is obese.

The basic statement concerns being obese. "In the US" narrows the scope from the entire world to just the US, and then we say whether we're talking about everyone in that scope, or just someone, at least one person. If two people are obese it is still true that "someone is obese".

As usual, the way we turn this into formal language is a bit tricky, because we need something that is more rigid than our fluid, flexible, spoken language. Formally these two types of statement would be rendered using "for all" and "there exists" like this:

> For all people X in the US, X is obese.
> There exists a person X in the US such that X is obese.

This sounds terribly pedantic in normal language, but makes things easier to manipulate in formal mathematics. "For all" and "there exists" are called quantifiers in mathematics; they quantify the scope of our statement.

ROCKY'S EGGS

In the case of Rocky promising all his eggs to the rats, the "for all" clause was fulfilled because "all" happened to be zero. This can often seem like a bit of a cheat although the logic strictly holds. This is called vacuous truth, or a condition being vacuously satisfied. Consider this statement:

> All the elephants in the room have two heads.

This sounds (and is) a ridiculous statement, but it is certainly true of the room I'm in at the moment. Unless you're reading this at the zoo, it's probably true of the room you're in as well. There are no elephants in the room, and all zero of them have two heads. This is related to the fact that a falsehood implies anything, logically. You might meet someone who very implausibly claims to be a billionnaire, and you might exclaim "If you're a billionnaire then I'm the Queen of Sheba!" In effect, this means you are completely sure the person in question is not a billionnaire. If a falsehood is true then truth and falsehood have become the same thing, which means everything is true, but also everything is false. It's not a very useful situation to be in.

NOT EVERYONE IS HORRIBLE

We can now revisit our emotional ouburst, "All men are sexist pigs!" This is technically a "for all" statement:

For all X in the set of men, X is a sexist pig.

so to refute it you have to show that there exists someone in the set of men who is not a sexist pig. So this is the negation of the above statement:

There exists X in the set of men, such that X is not a sexist pig.

This means you just have to find one man who is not a sexist pig. My friend Greg is definitely not a sexist pig. (Admittedly I can't really prove that without introducing you to him.)

One of my favorite mathematical jokes goes like this:

Three logicians walk into a bar. The bartender says "Would everyone like a beer?" The first logician says "I don't know." The second logician says "I don't know." The third logician says "Yes."

The point is that the bartender has asked a "for all" question, and the three logicians, being logicians, know how to verify and refute it properly. Either:

- *A*: All three logicians want a beer, or
- not *A*: There exists a logician who does not want a beer.

The first logician answers that they don't know, which means they definitely want a beer: otherwise they would know that there exists a logician who does not want a beer.

Likewise the second logicial must definitely want a beer, otherwise they in turn would know that there exists a logician who does not want a beer. The third logician can then answer on behalf of "everyone" because they also want a beer.

Like in many mathematical jokes, there is an element of truth in this that I find slightly endearing. I have spent enough time around mathematicians to know that this sort of precision is likely to spill over into their normal lives, where normal people would just answer "Yes" if they want a beer, even though that's not the technically correct answer to the question. Is it pedantic? It is in fact illuminating to the logicians, so perhaps it is still just narrowly on the side of precision.

ALL MATHEMATICIANS ARE AWKWARD

Sweeping statements are very close to stereotypes, and so are dangerous if you aren't open to the possibility of counter-examples, or if you don't respond to the situation at hand for what it is, rather than for what the sweeping statement says it is. I often complain about portrayals of mathematicians in popular culture, because they are too often awkward male people who are not very good at social interaction and are possibly insane. Someone recently said to me, "Yes, but mathematicians *are* like that!" This sounded awfully like "All mathematicians are awkward" and I didn't like that at all, because I am a mathematician and I believe I am not awkward. Therefore my existence refutes the statement:

For all X in the set of mathematicians, X is awkward.

If someone says to me "All mathematicians are awkward" after meeting me it sounds to me that they are implying either that I'm not a mathematician, or that I am awkward, because those are the only ways to reconcile my existence with the implication:

Being a mathematician implies being awkward.

Either they think I am a mathematician, therefore they must think I'm awkward, in which case I take offence. Or else they don't think I'm awkward, in which case they must think I'm not a mathematician, in which case I also take offence. There is one more possibility, which is that they think I'm not a mathematician but still awkward: double offence.

This sounds like an overanalysis, but that is like the difference between pedantry and precision. What's the difference between analysis and overanalysis? Sometimes people say to me "You're overthinking!" and I often want to reply "No, you're underthinking!" I think the difference is illumination: I don't call it overanalysis if it has helped me with something. In this case I find it helpful to know exactly why it is so frustrating when people make those sweeping statements to me about mathematicians.

NON-EXISTENCE

Imagine I say "Every female science student has been hit on by their supervisor." I actually heard someone say this at a panel event for women in science. Suppose you want to point out that this isn't true. What do you need to do? You just need to find one female science student who was not hit on by their supervisor, for example, me.

Whereas if I say "Some female science students have been hit on by their supervisor" and you want to say this isn't true, you have to do something much harder – you have to check every single female science student, and make sure that *nobody* has been hit on by their supervisor. Unfortunately this won't be possible.

In the first case you are trying to negate a "for all" statement, and in the second case you are negating a "there exists" statement.

We have the following negations:

1. World: All female science students in time.
   Original statement: For all female science students X, X has been hit on by their supervisor.
   Negation: There exists a female science student X such that X was not hit on by their supervisor.

2. World: All female science students in time.
   Original statement: There exists a female science student X such that X has been hit on by their supervisor.
   Negation: For all female science students X, X has not been hit on by their supervisor.

Just like "and" and "or", these two quantifiers go hand in hand as related by the negations: when you negate a statement involving one you get a statement involving the other, just like with "and" and "or".

If we add quantifiers to our logical language we now have what is called predicate logic, or first-order logic. The word "predicate" is to distinguish from "propositional", which is what we had without the quantifiers. "First-order" is to distinguish it from higher-order versions of logic, which are more complicated in the way that the quantifiers work.[1]

---

[1] Basic quantifiers only quantify over sets, that is, you can only say "For all objects in a certain set". You can't quantify over sets of objects, as that is a higher-order level of expressivity and would cause problems of self-reference. The difference is a bit technical, but the idea is one that we'll come back to when discussing paradoxes in Chapter 9.

YOU CAN ALWAYS BE RIGHT

If you are very precise about how you quantify your statements you can ensure that you are never wrong about anything. This is one of the reasons I as a mathematician might be quite annoying to argue with: I am careful to use enough quantifiers that it's almost impossible for me to be wrong. We've seen a few ways of doing this with phrases like

> In my opinion . . .
> In my experience . . .

Maybe we could add some more, like

> Maybe . . .
> Sometimes . . .
> Apparently . . .
> It seems to me . . .

I recently did an interview in which I lamented that some math lessons leave little lasting effect on people except math phobia – the students don't remember much actual math, and mostly only remember fear. In that case, teaching them math has been a waste of time and money, and worse, has actually had a negative effect. Thus we might have achieved a better result by teaching them no math at all, because that would have been a zero effect, which is better than negative. Unfortunately various tweets went out declaring that I said "Teaching math is a waste of time and money" and "We would be better off not teaching math at all." In fact, what I said is that *in some cases* teaching math is *something of* a waste of time and money, and in those cases, we might be better off not teaching it at all. Qualifiers abound.

My wonderful, wise, precise and illuminating PhD supervisor, Martin Hyland, is well known among his students for prefacing statements with "There is a sense in which". "There is a sense in which Mozart is more boring than Brahms" is another way of correcting my sweeping statement "Mozart is more boring than Brahms". "There is a sense in which teaching math can be a waste of time and money." The phrase has a wonderful way of focusing one's attention on exactly what is the sense – or what are the possible senses – in which something is true. It reminds us all that mathematics is not just about finding the right answer, but is about finding the sense in which things might or might not be true.

I believe that a useful way to be a rational person is to look for the sense in which things are true rather than simply deciding if they are true or false. Someone might say something that is untrue in strictly logical terms, but perhaps they were really trying to say something else, perhaps something with strong emotional content that we should listen to if we are intelligent humans rather than intelligent emotionless robots.

PART II

THE LIMITS OF LOGIC

8

TRUTH AND HUMANS

HOW TRUTH IS ACCESSED, CONVEYED AND RECEIVED

WE HAVE SEEN THE power of logic in producing rigorous unambiguous justifications. Now we are going to address the limits of that power. It is important to acknowledge those limits and not pretend that logic is the ultimate answer to everything: it certainly isn't.

When you discover your bicycle can't fly, should you throw it away? No. A bicycle is a wonderful device, as long as you don't try to use it beyond its limits, or beyond *your* limits with it. Cycling on the motorway or up Mount Everest might not go very well. Cycling in traffic would be a terrible idea for me but is a great way for better cyclists than me to get around. Sometimes in the rush hour you can even go faster than the cars. It does take more effort, but you might regard that as a good thing if you like being fit, or like burning up your own body fat as fuel rather than petrol.

Logic also has limits, especially in this messy, human, beautiful real world of ours. That doesn't mean it has failed, and it doesn't mean we should give up on it in some situations. But it does mean we shouldn't push it further than its limits. Rather, we should understand those limits and understand what we can do when we are beyond the reaches of pure logic. Understanding how and why logic has limits is the subject of this second part of the book.

I'm going to start by discussing something a little uncomfortable: the extent to which even mathematical proof is, in fact, a social construct, and so logical justifications in life are doomed to be as well. This might seem to fly in the face of everything

I've said about mathematics being completely rooted in logic, but the situation is more subtle than that.

In Chapter 2 we started discussing the fact that logic has no beginning – it has to start somewhere, and the starting point has to be some truths we assume, without justification. These are called axioms, and they are one aspect of the limits of logic. In Chapter 11 we will discuss how we come up with those axioms, as it must be by some means other than logic.

But another limit to logic is the end – when do we *stop* justifying things? Mathematical proofs are entirely *rooted* in logic, and they certainly don't go against logic. It's just that strictly logical proofs are impossible to write down for two reasons. One is that the logic they use depends on the rules of logic, and where did those come from? We have to assume some rules of logic even to use logic in the first place. We will come back to this paradox in the next chapter.

The other problem with writing strictly logical proofs, even once we've agreed to accept the basic rules of logic, is that they are utterly impractical to write down in full rigor, beyond a certain (rather low) level of complexity. And they would also not be illuminating even if we could do it. So what do we do instead? I think we can look at what mathematicians do to convince each other of their arguments, and extend those ideas to learn about how we can and should justify ourselves to other human beings in the wider world. Logic by itself is not enough. We do something a bit like trial by jury.

TRIAL BY JURY

Logic reaches its limits when it isn't powerful enough. One way in which that happens is when we simply don't have enough information or time, and then we have to resort to something other than logic to reach a conclusion. This is the subject of Chapter 10. But another way in which logic isn't powerful enough is when we need to convince someone else of our argument. Logic turns out to be a good way to verify truth, but this is not the same as convincing others of truth. Verifying truth and conveying truth are two different things.

Mathematicians set out to convince each other that a strictly logical proof is possible. A long proof is difficult to think up in very small steps, and so we usually sketch it out in broad brush strokes first, to see if the overall idea of the argument is likely to work. This isn't necessary if the argument is very short. If you're writing a quick email to someone you are very unlikely to plan it first – you just sit down, start writing it, say all the things you needed to say, and then sign off. But if you're writing a whole book, it would be quite extraordinary to simply start at the beginning and keep writing until you get to the end. In writing this one I started with broad ideas for the three parts, then ideas for chapter topics, and then ideas for the sections within chapters, and then a list of the main points in each section. It's sort of a fractal approach.

A fractal is a mathematical object that resembles itself at all scales, so that if you zoom in on a small part of it, that small part looks like the whole thing. In order for that to work, there has to be an infinite amount of detail, otherwise at some point you'll zoom in and there will be nothing more going on. This is a type of symmetry called "self-similarity".

Here is a picture of a fractal tree. At each level, each branch splits into two. And then at the next level, each branch splits into two again. This keeps going "forever". If you zoom in on any particular branching point, the part above it will look like a copy of the entire tree.

This tree represents how finding a proof works, in my head. The base is the thing you're trying to show is true, a bit like in our diagrams of causation in Chapter 5. The two branches going into it are the main factors that logically imply it. (Of course, there might be more than two main factors, and we can indeed have a fractal tree with more than two branchings at each point, but it becomes very difficult to draw so I'll just stick with two here.)

Next, you think about each of those main factors, and think about what are the main factors causing those to be true. So we get the next level of branches:

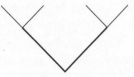

Note that there is still quite a large gap in between them.

So we think of each of those four factors, and what is causing them to be true, and if we keep going a few more times we get this:

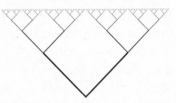

At this point the branches have become close enough together that there aren't really any gaps, and also, they are so small at the top that they are barely visible. Branching any further would hardly be distinguishable, so even though it's not a genuinely infinite fractal, we might as well stop from the point of view of showing another human what the fractal looks like.

This is somewhat how proofs work – at some point you just decide to stop filling in gaps any more, because you think that justifying things any further would not help. In arguments in real life, we should keep going until the other person is convinced, or until we realize that our fundamental starting points are so different that we will never be able to convince them unless we can change their fundamental beliefs.

In practice mathematical proof is a bit like trial by jury. In experimental science, peer review means that some scientific peers decide whether they think they could reproduce the experiment you did. The peer scientists don't actually have to try to reproduce it, they just decide if they are convinced that they could. In mathematics, peer review means that some mathematical peers decide whether they think that the proof could be made entirely logical. They are unlikely to try turning it into a strictly formal logical proof, but they are quite likely to try filling in some of the gaps between branches, to see if they can. Disagreements can arise when a mathematician doesn't see how to fill a gap in, but in that case they ask the person who wrote the proof, and the onus is then on the original author to fill at least some of the gap in until the skeptical mathematician becomes convinced.

The point about trial by jury is that unless someone confesses to a crime (and even if they do), it is highly unlikely that definitive proof can be found that someone did it. So the burden of proof becomes not logical proof, but a sociological standard – you have to be able to convince a jury of peers, that is, randomly selected people. It is a flawed system but it is a clever one in unideal circumstances. It is flawed because the people on the jury are in fact people, so they are susceptible to emotions and confusion, and so the art of the trial lawyer might focus more on how to sway the emotions of those people more than presenting the logic of the situation.

Peer review is flawed in similar if different ways. The peers who are reviewing the proof are humans so they too can be swayed by emotions, such as their feeling about the author's reputation. You might think that papers should then be reviewed anonymously, but in practice this is impossible. It's a bit like marking exams anonymously if your class has only three students and you've been working with them closely all year – whether or not they write their names on the exam paper, you will know exactly which student is which. The same is true for research in very high-level mathematics – there aren't that many people working in each rarefied field, and good researchers are likely to present versions of their work at conferences first, to road test it a bit. In life as well, there are people we decide to trust, and we listen to what they say with a predisposition to believe them, whereas with other people we have a strong urge to be skeptical. This may or may not be justified, and we'll discuss this more later.

Mathematicians are sometimes reluctant to admit to these sociological and human aspects of math outside of mathematical circles, because they don't want to cast doubt on the truth of their work. We might not want to admit that our framework of rigorous proofs and peer review is anything less than completely, totally and utterly rock solid. But I believe that overstating the remit and achievement of a system is dangerous because it gives people an opportunity to doubt one part of what you say, and thus doubt all of it. I am dismayed when people dismiss mathematics as irrelevant or boring, but I am also dismayed when people place it on a pedestal of universal power and irreproachability. I would rather we appreciated mathematics for what it is: something in between. Something that helps us sort out our messy human world, but is necessarily still a little bit messy in places. Something powerful and relevant, but with limits.

The interaction between the logical and the human parts of the mathematical process can teach us about those interactions in all human discourse. We'll discuss this more in Chapter 15 on emotions.

Research papers also deliberately include material that is not strictly logical, to help mathematicians trying to understand the logical proofs. The help comes in the form of analogies, ideas, informal explanations, pictures, background discussion, small test-case examples, and more. None of this is part of the formal proof, but is part of the process of helping mathematicians get their intuition to match the logic of the proof. We know that if the logic is sound but the outcome doesn't match someone's intuition, they will still be skeptical. Dealing with skepticism is an important part of the mathematical process: we aim to iron out all possibilities for reasonable skepticism. This is like the idea of "reasonable doubt" in trials.

However, it is worth noting that there is a fine line here between reasonable and unreasonable objections on grounds of intuition. This is a bit like the difference between peer review and audience vote.

REASONABLE OBJECTIONS

Competitions decided by an audience vote are often somewhat derided, and "experts" might poke fun at non-expert audience members who, the experts say, don't know what they're talking about. This happens when audiences love crossover singers who sing a famous and heart-rending excerpt from an opera (say, *Nessun Dorma*), but without "proper opera singing technique". However, according to the rules of an audience vote, that stirring singer can win fairly and squarely. Unlike peer review it's just a vote, and nobody has to justify their vote. In the process of peer review, objections need to be justified too, not just stated.

In normal life, outside some specialized situations, there isn't a clearly defined panel acting as peer review on our logical arguments. In a court of law it's the jury, and their decisions might be based less in the logic of the arguments and more in their emotional response to witness testimony. Either way, jury is not generally required to justify its decision. For politicians, the "peer review" is the election – it doesn't matter if they're right or not, and it doesn't matter if their arguments are sound or not, it just matters if people vote for them or not. Voters do not have to justify their vote either. You might hope that the votes will depend on the soundness of the politicians' arguments, but probably not if you've ever experienced an election. For corporations, the "peer review" is the money: they just need to persuade people to buy their product, and it doesn't necessarily matter if their methods are sound or logical (although if they're outright fraudulent they might get into legal trouble).

Politicians, corporations, and anyone seeking to sway people's opinions can be seen to be using non-logical methods to influence and manipulate them via their emotions. It is easy to be swept along by this, but if you don't want to be so easily manipulated it is important to remain skeptical. This doesn't mean outright denying everything everyone says, but requiring at least some level of justification, and having a framework so that you are prepared to believe someone if they achieve that level, and not if they don't. This is the difference between reasonable and unreasonable skepticism, and we will come back to it when discussing how to be rational in Chapter 16.

Reasonable skepticism about a mathematical proof can arise in two ways:

1. Someone might think there's a gap or error in your logic.

2. Your conclusion might contradict someone's intuition.

The first is a simple logical objection, and is dealt with in a simple logical way, by filling in more of the logic to make clear that the gap can be filled in or that the supposed error isn't one.

The second type of objection is more slippery. It has happened to me several times in my research and happens all the time in politics. It happens every time someone doesn't believe some scientific research because it contradicts their own experience or strongly held belief. It's why some people still believe that vaccinations cause autism although there is no scientific evidence for it. It's why some people still believe the universe is only a few thousand years old or that the earth is flat or that human life did not originate in Africa or that Barack Obama was not born in Hawaii, despite evidence.

An intuitive objection is much harder to deal with than a logical objection, because you have to change someone's intuition in order to convince them, and there is not a clearly defined way of doing that. It should be clear that restating the evidence is not going to help, and telling people they are stupid definitely doesn't help. We will talk more about this in Chapter 15. In theory, this type of objection has no place in rigorous mathematics, because if someone can't find a fault in your proof then they do not have a valid objection. However, mathematical research is still about convincing humans of things, and so in practice this very human type of objection is a problem, not least because if humans aren't convinced of your result then they won't use it, build on it, or value it.

When this sort of objection happens to me I do console myself with the fact that logic contradicting intuition is one of the reasons we use logic at all – if logic always matched intuition it would be somewhat redundant to use it. Crucially, I don't just dismiss the intuitive objection, but rather, I try to find the root of it so that I can resolve it. Often something seems intuitive from one point of view but counter-intuitive from another, and so resolving the conflict involves persuading someone to acknowledge the other point of view. But it is also important to acknowledge the validity of their point of view. To do this, we have to start by understanding what their point of view is, to see where their intuition is coming from.

Whether we're writing mathematical papers and talks, or developing arguments to back up our own views, being able to imagine a skeptical person is an important discipline, so that we can pre-empt these situations and perhaps remove people's reasonable skepticism in advance.

I often imagine a skeptical person arguing with me. We are allowed to imagine that they are as intelligent as us, which is why it is called peer review and not idiot review, but I imagine that they are highly skeptical of everything I'm saying, or that they are actively trying to find a mistake in my proof, so that I can find any possible mistakes myself.

Imagining someone very skeptical arguing with you is a good way to test your logic in life as well. It does require you to be able to think like someone else, but this in itself is an important skill and a crucial aspect of having discussions that build bridges with other people rather than widening divides. It can also genuinely open your eyes to new ways of thinking about something. I find that I understand things much more by teaching them, because I have to think about how to explain them to skeptical students. Even while writing this book I keep having new revelations about the interaction between logic and the world.

If you only imagine talking to people who already agree with you then you never have to test your arguments. Worse, many people only talk to people who agree with them, both in reality *and* in their imagination. The online version of this is the much-discussed "echo chamber" that we are apparently being kept inside by search engines and social media algorithms. To avoid this, I think it's important to seek out different viewpoints to try and understand where they are coming from. I sometimes test out my powers by reading what I think is a reasonable article and then trying to guess what the main objections will be in the online comments section. Often the disagreements come from differences in fundamental beliefs, as we'll discuss in Chapter 11. But sometimes the comments are knee-jerk reactions that may well be true but have nothing much to do with the argument in the article. Whenever there is an article about dogs there is bound to be a thread of comments declaring, apropos of nothing in particular, that "they eat dogs in China". This brings us to the difference between truth and illumination.

TRUTH VS ILLUMINATION

They may well eat dog meat somewhere in China, but this doesn't make that statement relevant to an article about, say, a dog-walking app matching busy dog owners with freelance dog walkers in Chicago. Not all truth is relevant or helpful. Things that are true are not necessarily illuminating. This is another sense in which logic reaches its limits: truth can be assessed using logic, but illumination can't really. Truth and illumination should not be mistaken for each other but the way in which they interact is important. Again we can start looking at this through the lens of mathematical truth, and the surprising fact that *equations are all lies.*

Did you react to that assertion? I'm afraid I said it largely just to get a reaction. It was like clickbait, although there was nothing to click on. Admittedly the statement "Equations are all lies" is not true, but I said it to make a point. (Indeed, later we'll discuss the fact that truth and attention grabbing are largely independent.)

Here is an equation that is not a lie: $1 = 1$. However, that equation is true but not at all illuminating, so I don't really want to count it as an equation. So, as we learned in the first part of this book, I should really refine my statement. I could say "Most equations are lies" but that might not be true: there are after all infinitely many true equations of the form $x = x$ because there is one for every number at least:

$$1 = 1$$
$$2 = 2$$
$$3 = 3$$
$$4 = 4$$
$$\vdots$$

and so on. What I really mean is that the only equations that aren't lies are trivial, in which case they're pointless. So:

All equations that are illuminating are lies.

What do I really mean by this? As we mentioned right at the start, one of the enduring myths about mathematics is that it is all about numbers and equations. While this is not exactly true, numbers and equations certainly play a central role. So how can I say that all those equations are lies? Admittedly I'm still playing to emotions a bit by using the word "lie". What I really mean is that all equations are hiding something that is *not* an equality, so they are not really, logically, totally and utterly an equation. For example let's think about this equation:

$$10 + 1 = 1 + 10.$$

You might remember this as the commutative law for addition, or you might just know instinctively that if you take ten things and one thing you will have the same number of things as if you took one thing first and ten things afterwards. However, this is something that children have to get used to. I have helped children with math at school for years, and when they first learn to add up by "counting on", the commutativity of addition is not at all obvious. If you ask them ten plus one they will happily put ten in their head, count on one with their fingers, and get to eleven. But if you ask them one plus ten they will put one in their head and arduously count on ten with their fingers. Depending on how adept they are at it, they may or may not arrive at the correct number after all that laborious counting. To them, ten plus one is not the same process as one plus ten.

In fact, in high-level mathematics, $1 + 10$ is not defined to be the same thing as $10 + 1$. That is why the commutative law for addition is a law and not a definition. Also, that is why the equation above is useful and illuminating. It tells us that although ten plus one is a different process from one plus ten, they will produce the same answer, and so we can pick whichever one is easier for us. Once children work this out, they can use it to help them add up by counting on, knowing that it will always be easier to start with the larger number in their head and count on by the smaller number. So this equation is not a true equation: the left-hand side and the right-hand side are not exactly the same. Its power is in the fact that in one sense (the process) they are different but in another sense (the answer) they are the same.

All equations that we study in mathematics are like this. They show us two things that can be regarded as the same in some way even though they are different in other ways. This is how equations help us. If there was really nothing different about the two sides, the equation would be true but unilluminating. The equations that really have nothing different about the two sides are the ones of the form

$$x = x$$

and these ones are never useful.

HOW WE CONVINCE PEOPLE

We have seen that the logical truth is not always illuminating, and in practice it often isn't logic that actually convinces us of something. This is related to the fact that logic is not usually helpful for thinking up a proof in the first place. When we're thinking up a proof we often use gut instinct, vague suspicions, hunches, inklings, we look for things that slightly remind us of other things, we wait for flashes of inspiration. We then try to fill in all those things using logic, but only after we've used many not-entirely-logical processes to get our first ideas in place. This is perhaps the origin of the myth of math "geniuses" – there is often a mysterious element of inspiration at the start of mathematical work, but let's not forget the sheer hard work that goes into constructing the logic afterwards.

We don't entirely use logic to *understand* logical proofs either. Often in a research paper a logical proof will be accompanied by a description of what "the idea is", which is something more informal, not rigorous, but invokes ideas and imagery that might help us to understand the logic. It might be a picture, like the fractal tree picture I used to talk about filling in the gaps in a proof. It might be some sort of schematic diagram showing how things fit together, like my diagrams of causation in Chapter 5. It might be an analogy, or a small example. These things are not exactly logical themselves but they help us *understand* the logic. Often once we've understood the idea like this we can fill in the logical steps ourselves with much less help. *Feeling* why something is true helps us understand why it's true *logically*, even in an area as abstract as mathematics. If we don't get the idea of why it's true we might follow every step of the logic but still feel disconcerted that we don't feel like we *really* understand what is going on.

The experience of knowing something logically but not emotionally happens in proofs, and it happens in grief, when you know intellectually that something terrible has happened but some emotional part of you can't yet believe or accept it. I think this is the difference between knowing something intellectually and knowing something emotionally. Our intellect and our emotions don't necessarily take us to different conclusions all the time, it's just that sometimes there's a lag between them.

This comes into play when we are learning things. If we engage our emotions or personal experiences while learning something it is quite likely to get embedded more deeply in our consciousness. People say that the best way to learn things is by experience, and I think this is because if we learn something by experience we really feel what it's like, and then the thing we've learned sits somewhere deeper inside us than if we've just read about it in books.

This is related to the contentious argument about the role of memorization in learning math. Some people take it entirely for granted that some rote memorization is necessary if you're going to be good at math. But other people, often professional research mathematicians themselves (including me), are convinced that they have never really memorized anything in math.

In fact, one of the main reasons I always loved math was exactly the fact that it didn't require memorization, only understanding. And yet, many people tell me the whole reason they were put off math was having to memorize things. I sympathize, as I am also put off by having to memorize facts, I just don't agree that math requires that. Usually when I say this someone will challenge me and say "But surely you learned your times tables." Somehow times tables always come up as the essential thing that surely nobody could avoid memorizing.

Now, I am certainly not an amazing whizz at arithmetic, not one of those human calculator types who enjoy rapidly multiplying five-digit numbers in their head. However, I am perfectly fine at basic arithmetic and certainly above average compared with the general population, and yet I have never memorized my times tables. I *know* my times tables by some other, more subtle route that does not involve memorizing. I think this is like the fact that I know my name, but I have not *memorized* my name. I have internalized it.

Logic, math and science can be difficult to internalize if they appear to be devoid of emotional content. We should be clear that it is only the methods of justification that should be devoid of emotional content, not the methods of communication and understanding. Emotional engagement is a much more powerful way to make logical truth convincing. In fact, it can make anything convincing, logical or otherwise, true or otherwise, as can be seen from the success of so-called internet "memes".

MEMES

Things can be true without being illuminating, but things can also be illuminating without being true. Internet memes are a rich source of this latter type of situation.

I saw one this week on the subject of the scientific method. The text said this:

HOW IT SHOULD BE:

Scientists (experts): There's a problem.

Politicians (non-experts): Let's debate a solution.

HOW IT ACTUALLY IS:

Scientists (experts): There's a problem.

Politicians (non-experts): Let's debate if there's a problem.

On the whole I think the point it's trying to make is that the interaction between science and politics has become unideal, with politicians overstepping their remit and encroaching on tasks that should be dealt with by scientists. However, I disagree with the summary of "How it should be", because I think scientists should actually say something more like "We are 99 percent sure there's a problem" and then scientists should investigate what the solutions are. Politicians should then debate whether or not the solutions should be funded, which should come down to weighing up the dangers involved with not solving the problem, the level of certainty of the scientists, and the cost and likelihood of success in solving the problem. However, this is much less catchy and probably too much text to fit in a meme.

Another example I saw recently said:

Funny how no country has ever
tried to repeal universal healthcare.

It's almost like it works.

Again, I agree with the overall point which is, I believe, to support universal healthcare. However, I'm not sure if the meme is true – arguably some people have been trying to destroy and privatize the National Health Service in the UK.

However, details, nuance and well-structured arguments do not help memes go viral. Rather, catchy phrases and pithy one-liners help memes go viral (together with amusing graphics).

This sometimes makes rational people throw up their arms in despair. But I think we can do something better than that: we can learn from it. Explaining something to someone who doesn't already understand it always involves making it digestible in some way. With some advanced mathematical or scientific topics that means simplifying it for people who haven't gone through the years of training necessary to understand it fully. Some people think this compromises the subject so much that it's not worth it, and so they don't bother with any attempts at "popularization" of science.

However, I disagree with that. I think that we can find ways of simplifying arguments that still capture their essence, and also attempt to engage emotions and amusement in the way that memes do. When I saw the above meme about science, I wanted to edit it. I wanted to find a trade-off between the full nuanced description, which is probably too long, and something which is still punchy but more accurate, perhaps this:

HOW IT SHOULD BE:

Scientists (experts): We think there's a problem. Here's a solution.

Politicians (non-experts): Let's debate funding their solution.

HOW IT ACTUALLY IS:

Scientists (experts): We think there's a problem.

Politicians (non-experts): Let's debate if there's a problem.

In summary, we should look at engaging people's emotions to convince them of logical arguments, rather than using logical argument alone. In the rest of this book we are going to look at various ways in which logic has limits, and how emotions can help us get beyond those limits. We should particularly not pit emotions against logic. They are not opposites, but can work together to make things that are both defensible and believable.

9

PARADOXES

WHEN LOGIC CAUSES CONTRADICTIONS

I AM AN AVID writer of to do lists. I find it an excellent way to procrastinate in a mildly useful way. Sometimes if I'm feeling particularly tired or stressed I will put some very easy things on my to do list so that I can easily declare that I've achieved something. This might be something like "eat breakfast" or "check email". I went through a phase of putting "get up" on my to do list, so that I could immediately cross that off. It suddenly occurred to me to put "Do something on this list" on my list. Could I then immediately cross that off? I got quite confused wondering about that.

I often get intrigued by looped-up thoughts like this. For example, I really want to go round telling everyone not to give unsolicited advice, but I fear that this would constitute unsolicited advice. Then there's the fact that I can't keep ice cream in the house because if there is any, I will immediately eat all of it, and then there won't be any. A more serious contradiction happened to me when I was filling in a form online for a visa application. It sternly told me that I had to fill in my full name and it had to match my name in my passport. However, my middle name in my passport has a hyphen, and the online form would not allow me to input a hyphen. "Invalid name: only alphabetic characters are allowed." So I found myself stuck – I had to fill in my name exactly as it appeared in my passport, but was not allowed to. I am surely not the only person with a hyphenated name ever to have applied for a visa, not to mention those with apostrophes or accents.

I think of these loops and contradictions as paradoxes of life.

Paradoxes occur when logic contradicts itself or when logic contradicts intuition. Both cases show us some of the limitations of strictly logical thinking. In the first case we see how we may need to take more care over how we set up our logic, our definitions, or the scope of our thinking. In the second case we see that we shouldn't necessarily trust our intuition, or that we should take some time to understand quite where that intuition is coming from. The first type illuminates our treatment of logic and the second illuminates our view of the world.

In this chapter we'll explore some of my favorite mind-bending paradoxes, some of which are famous mathematical paradoxes, and some of which are weird things I've noticed about life. Historically, paradoxes have sometimes been so illuminating that they have led to huge developments of entire fields of mathematics. They are a very interesting place to study the limits of logic. They are curious situations where pursuing logic too hard appears to cause contradictions.

THE LIAR PARADOX

It is remarkably easy to get yourself into a logical paradox: just declare "I'm lying!" I have seen logically inclined children do this and then collapse into giggles. They have created a paradox – if they're telling the truth, then they're lying. But if they're lying, they're telling the truth. This is a classical paradox known as the liar paradox. It also happens if I say "Don't take my advice." Should you then take that advice? If you take it, that means you shouldn't take it. And if you don't take it, that means you've taken it.

We can make a pair of statements into similarly loopy paradoxes like this:

1. The following statement is true.
2. The previous statement is false.

There is also the curious sentence discussed by Douglas Hofstadter in *Metamagical Themas*:

Cette phrase en français est difficile à traduire en anglais.

which can be translated literally as

This sentence in French is difficult to translate into English.

but it no longer makes sense.

At this point I should apologize to the French translator of this book, if there is one. Except of course, that sentence has just made it worse because it is somewhat difficult to translate into French.

CARROLL'S PARADOX

Lewis Carroll is probably best known for being the author of *Alice's Adventures in Wonderland* but he was in fact a mathematician called Charles Lutwidge Dodgson who worked at the University of Oxford. He wrote about a logical paradox in an article entitled "What the Tortoise Said to Achilles" for the philsophical journal *Mind*. The article is written in the form of a dialogue, in which a tortoise argues Achilles into a corner, or rather, an infinite abyss. His use of the tortoise and Achilles characters is a nod to Zeno's paradoxes, which we'll look at shortly.

In Carroll's paradox, he explores how logical arguments are built up by logical implications. We discussed in the previous chapter the fact that logic reaches a limit because in order to use it we have to start by assuming some rules of logic. Carroll's paradox is about what happens if we don't do that – we just never get there, endlessly filling in more logical steps and never reaching a conclusion. It's like trying to draw a genuinely infinite fractal tree, rather than one that just looks filled in enough – you can never finish in a lifetime.

Carroll's tortoise asks Achilles to compare two sides of a

triangle to see if they are the same length. You might pick up a ruler and measure both sides and discover they are both 5 cm and so they're the same as each other.

This involves the statements:

A: Both sides of the triangle equal the length of 5 cm.

Z: Both sides of the triangle equal each other.

The tortoise asks Achilles if Z follows from A, and Achilles says yes of course. Perhaps you agree. But the tortoise says this is only true if we know that A implies Z. So we have another statement mediating between them:

A: Both sides of the triangle equal the length of 5cm.

B: A implies Z.

Z: Both sides of the triangle equal each other.

Now the tortoise asks Achilles if A and B together imply Z. But that is itself a new statement:

A: Both sides of the triangle equal the length of 5 cm.

B: A implies Z.

C: A and B imply Z.

Z: Both sides of the triangle equal each other.

Now the tortoise asks if A, B and C imply Z. But that is a new statement:

A: Both sides of the triangle equal the length of 5 cm.

B: A implies Z.

C: A and B imply Z.

D: A, B and C together imply Z.

Z: Both sides of the triangle equal each other.

The story ends with the tortoise still sitting there torturing Achilles by making him write down all these intermediate statements; it becomes clear that this will go on forever.

So how do we ever deduce anything from anything else? Or have we in fact never correctly deduced anything from anything else? The answer is that we have to use the rule of inference, modus ponens (mentioned in Chapter 4), which we must assume to be valid in order to get anywhere at all. This paradox warns us that there is always a level of meta-logic controlling our logic, and we can only understand this by keeping those levels separate.

This is somewhat like attempts to find all the factors that caused something to happen, as we discussed in Chapter 5:

A: I dropped the glass.

B: A implies Z because the glass was too fragile.

C: A and B imply Z because the floor was too hard.

D: A, B and C together imply Z because gravity intervened.

E: A, B, C and D together imply Z because I didn't catch the glass.

F: A, B, C, D and E together imply Z because nobody else caught the glass.

G: A, B, C, D, E and F together imply Z because. . .
$$\vdots$$

Z: The glass broke.

If we don't decide enough is enough, we will never be able to conclude anything.

ZENO'S PARADOXES

Lewis Carroll's use of the tortoise and Achilles is a tribute going a couple of thousand years back to Zeno, who used these characters in a different, more concrete paradox. He imagines a tortoise in a foot race against the extremely fast Achilles, in which the tortoise gets a head start. But then Zeno argues like this: by the time Achilles gets to the place where the tortoise started, the tortoise will have moved forwards a bit, say to point B. By the time Achilles gets to point B, the tortoise will have moved forward a bit, say to point C. By the time Achilles gets to point C, the tortoise will have moved forward a bit, say to point D. This will go on forever and so Achilles will never overtake the tortoise. And yet, in reality, we know that Achilles will definitely win the race.

Zeno was a Greek philosopher who lived in the fifth century BC, and several famous paradoxes are attributed to him. Like many paradoxes over history, these ones are arrived at through thought experiments, with the aim of trying to understand some fundamental aspect of how to study the world. This is different from trying to understand the world.

The three most famous of Zeno's paradoxes are to do with motion, distances and infinitely small things. The first is the one about the tortoise and Achilles. The next one involves travelling from A to B all by yourself. Zeno argues that first you must cover half the distance, then half the remaining distance, then half the remaining distance, and so on. This will go on forever, so you will never get there. And yet, in reality, we do succeed in travelling to places every day.

The third paradox involves an arrow flying through the air. Zeno argues that if you only saw it for one instant, you would not see it moving. This is true at any instant at all. Therefore how can it be moving? And yet we do know that things move although, indeed, at any given instant nothing can be seen to be moving. This is why photos are stationary images, unlike videos.

Paradoxes fall broadly into two types. The first is *veridical paradoxes*, in which there is nothing wrong with the logic, but the logic pushes us into a situation that is at odds with our view of the world. The second is *falsidical paradoxes*, where a fault of logic has been hidden in the argument, and that is what causes the strange result.

Zeno's paradoxes are falsidical paradoxes: the error is in the logic, not in our intuition about the world. The error is very subtle though, and it took mathematicians a couple of thousand years to work out how to correct it. It all comes down to how you interpret "forever", and how you think about sticking instantaneous moments together to make longer periods of time. In fact, it all comes down to how we deal with the question of sticking together infinitely many infinitesimally small things. If you do it without nuance, or assume it works the same way as adding together finitely many finite things, you produce these strange paradoxes. What this warns us is not that our view of the world is wrong, but that we need to be more careful about both infinitely large and infinitely small things.

Taking care over infinitely small things leads to questions about sliding scales and gray areas, which we mentioned in Chapter 4 and will come back to in more depth in Chapter 12 on gray areas. Taking care over infinitely large things includes questions about when it is valid to add up an infinitely long string of numbers. A popular Numberphile video claimed that adding up all the numbers 1, 2, 3, and so on forever "equals" $-\frac{1}{12}$. They argued using what they call "mathematical hocus pocus" which actually consisted of some leaps of logic that were perhaps intuitively meaningful but logically flawed – they made unfounded assumptions about how infinitely long strings of numbers behave.

I hope you feel that the end result is absurd, not least because all the numbers we're adding get bigger and bigger to infinity. Indeed this is why the infinite sum

$$1 + 2 + 3 + \cdots$$

cannot be said to have a sensible answer without substantial qualification. There is some very profound mathematics that gives *a sense in which* this "equation" has a meaning, but it is definitely not by adding up this infinite quantity of numbers.

Unfortunately the video fooled millions of people, partly because of the good reputation of Numberphile videos in general. It is perhaps a case in point about memes being popular and believable even if they contradict both logic *and intuition*. It is perhaps also an example of a general belief that math is a bit ridiculous, which is very unfortunate. What the "equation" should be is a starting point for thinking about how infinitely large things cause strange situations, as in the next example.

HILBERT'S PARADOX

The paradox of Hilbert's Hotel is a thought experiment about infinitely large things causing peculiar situations.

David Hilbert was a mathematician who lived almost two thousand years after Zeno, but mathematicians were still (and are still) trying to understand infinity. Hilbert's thought experiment involves an infinite hotel, with rooms numbered 1, 2, 3, 4, and so on forever. Imagine that the hotel is full, so you also need to imagine an infinite number of people. Neither the infinite hotel nor the infinite people are possible in real life, but this is a thought experiment. Now imagine that a new guest arrives. The hotel is full, so no room is available for the new guest. However, we could move everyone up a room, so that the person in room 1 moves to room 2, the person in room 2 moves to room 3, and so on. Because there are an infinite number of rooms, everyone has a room they can move to, at the slight expense of kicking the occupant out. But after being kicked out, that occupant can in turn move to a new room. All this leaves room 1 empty and the new guest can move in.

The paradox here is not in the logic but in our intuition. In normal life, if a hotel has no empty rooms you can't just move people around and miraculously make an empty room appear, without getting people to share rooms. The difference is that in normal life all hotels are finite. This is a veridical paradox that challenges our intuition around infinity. It warns us that we can't just extend our intuition about finite numbers to infinite numbers, because strange things start happening. Those things aren't wrong, they're just different.

Hilbert's Hotel paradox can be extended to thinking about more new guests arriving, and even infinite new guests arriving. It leads to the study of infinity as a new type of number that does not obey the same rules as ordinary numbers.

This might seem rather removed from real life as we do not have infinitely many of anything in real life. Or do we? One way we consider having infinitely many things in life goes back to Zeno's paradox, and in the fact that any distance can be divided into infinitely many increasingly small distances. This might seem like a technicality, but remarkably this technicality enables us to understand motion and is therefore critical to everything that is automated in our modern world.

But another way we have effectively infinitely many things is by thinking about unlimited supply. Hilbert's hotel has an unlimited supply of hotel rooms, and another vacant room can always be produced with essentially no incremental cost. This is somewhat like the situation with digital media now, as extra copies of files can be made at will, with no incremental cost. Although we do not actually have infinitely many copies of a file, it makes some sense to model the situation as having an infinite supply, which goes some way to explaining why the value of digital media has plummeted to zero, or very close to it. This is of course related to the issue of piracy, and some people argue that the supply is only infinite because of pirates, and thus that preventing people from stealing digital content will cause the supply to go back to being finite. Another point of view says that copying digital content is not really the same as "stealing", because you're not removing the object from someone. Indeed, the theory of infinity developed following Hilbert's paradox tells us that subtracting one from infinity still leaves infinity. The math can't tell us what to do about these moral issues, but it can give us clearer terms in which to discuss them.

GÖDEL'S PARADOX

All of these paradoxes are related to, and historically lead up to, Gödel's incompleteness theorems. Kurt Gödel was a logician who lived from 1906 to 1978. In 1931 he proved a theorem about the limitations of mathematics which proved quite shocking to mathematicians of the time. The theorem basically says that any consistent logical system is doomed to have statements that can neither be proved nor disproved, unless the logical system is rather small and boring. "Consistent" and "logical" have formal meanings here: logical means it has been built up from axioms in a precise way, and consistent means that it does not contain any contradictions, so that if something is true it can't also be false.

Of course, "small" and "boring" are very informal words, and sound like subjective descriptions. But, for example, any logical system that is merely large and interesting enough to express the arithmetic of the whole numbers is already doomed to have this property of incompleteness. First-order logic, without quantifiers, doesn't fall into this category. In fact, first-order logic can be proved to be complete, in that everything is provably true or false. Second-order logic cannot.

A bit like Russell's paradox (see the next section), it comes down to questions of self-reference. As soon as a statement is allowed to reference itself, strange loops can be caused. Sometimes these loops produce beautiful structures like fractals or infinite loops in computer programs. But logical loops can cause us problems, as referenced in Douglas Hofstadter's book title *I Am a Strange Loop*.

Gödel's incompleteness theorem is discussed extensively in Hofstadter's exquisite earlier book *Gödel, Escher, Bach*. In it Hofstadter elucidates not only the incompleteness theorem but all sorts of fascinating links between logical structures and abstract structures in the music of Bach and the prints of Escher, both of whose works are deeply mathematical while also being immensely artistically satisfying.

The proof of the incompleteness theorem involves constructing a statement that creates a paradox through self-reference. The ingenuity and shock of it comes from being able to do this entirely formally in a mathematical system, essentially using numbers. It is easy enough to utter a sentence in normal English that is unprovable, such as "I am happy", but that is just because "happy" is not a logical enough concept to prove or disprove using logic.

Before Gödel's theorem, many mathematicians believed that, unlike the real world, the mathematical world was a perfectly logical world in which everything was provable. Gödel threw cold water all over that. He basically formally encoded the sentence.

This statement is unprovable.

First of all we can determine that this statement is true: if it were false then this would mean it is provable, but that would make it true and would give a contradiction. However, the fact that it is true means that it is unprovable, because that is what the statement says. (If you're anything like me, you might get dizzy thinking about that.)

Gödel showed that it is possible to make this statement using the language of arithmetic, thus showing that any mathematical system that includes arithmetic must be incomplete. There certainly are smaller mathematical systems than that which are complete, but they don't include arithmetic so can hardly claim to be all of mathematics.

Gödel's paradox is a veridical paradox – there is nothing wrong with the logic, although some mathematicians were so outraged by its conclusion that they refused to believe it. This is an example of the fact that even in the logical world of mathematics if a conclusion feels wrong there are mathematicians who refuse to believe it although they can't find anything logically wrong with the proof. What the paradox warns us is that we should limit our expectations of what mathematics can do. Mathematicians have by and large recovered from that shock now.

In fact, even before this shock there was an earlier threat to the very foundations of mathematics, coming from Bertrand Russell.

RUSSELL'S PARADOX

When I meet people and say I'm a mathematician I often get slightly strange responses. The usual ones are "Ooh, I'm no good at math" or, more recently, "I wish I understood more about math". It's funny how some people immediately boast about how bad they are at math, but other people immediately try to show off how knowledgeable they are. I once met a guy at a wedding who immediately said "Doesn't Russell's paradox show that math is a failure?" This was a particularly curious approach because someone who knows enough about math to know this thing called Russell's paradox would usually know enough to understand why this doesn't mean math is a failure. But of course he was the kind of guy who wanted to belittle me, probably because he felt inadequate.

Bertrand Russell (1872–1970) was a philosopher and mathematician (among other things). His paradox dates from 1901 and is also related to questions about keeping levels separate. It can be stated in informal terms like this.

Imagine a male barber in a town. The barber shaves all the men in the town who do not shave themselves, and nobody else. Who shaves the barber?

Now, the barber is supposed to shave all the men who do not shave themselves. So if the barber does not shave himself, then the barber is supposed to shave himself, and if he doesn't shave himself, then he does. This is very tangled up. Let's consider any man A in the town:

- If person A shaves person A, then the barber doesn't shave person A.

- If person A does not shave person A, the barber shaves person A.

This results in a problem if person A *is* the barber, which is allowed because A represented any person in the town. In this case the two statements become:

- If the barber shaves the barber, then the barber doesn't shave the barber.

- If the barber doesn't shave the barber, then the barber shaves the barber.

Each of these statements produces a contradiction. This is Russell's paradox. Formally, the paradox is stated in terms of sets instead. It says: consider the set S of all sets that are not members of themselves. Is this set a member of itself? If it is, then it isn't. And if it isn't, then it is. It's a paradox.

The problem in this situation is not the logic per se but the very statement we started with. With the barber, we simply conclude that no such barber can exist. We have to do the same for the sets as well: no such set S can exist, which is to say that this is not a valid way of defining a set.

Russell's paradox did not break the entirety of mathematics, contrary to what that guy I met at the wedding tried to claim. Rather, it highlights an important nuance that needs to be taken into account when we're defining mathematical sets, which is that some descriptions allow sets that will result in a contradiction, so we have to be careful to rule that possibility out. This led to the very careful axiomatization of set theory so that a set is not just "a collection of things" but is "a collection of things that can be defined by a particular list of constructions, and no others". The technical goal of the axioms is basically to avoid Russell's paradox. The idea is to say that we have different "levels" of sets, a bit like how we have different "levels" of logic.

Russell's paradox is caused by statements involving sets that loop back on themselves. As long as we have different levels, we can declare that a set of all sets is on a different level, and this prevents us making statements that loop back on themselves.[1]

This paradox, like many studied by mathematicians, might seem rather technical and abstract. However, it has helped me understand some crucial questions about tolerance in society.

TOLERANCE

I sometimes see people getting very looped up thinking about tolerance and open-mindedness. You might aspire to be a tolerant and open-minded human being, and I would argue that this is a good thing. But does that mean you have to tolerate hateful and intolerant views? Does that mean you have to be open-minded towards closed-minded behavior? I would argue that you do not. I think this is a subtle form of Russell's paradox, and that we can resolve it by narrowing the scope of our quantifier. We might think that tolerant means "tolerant of all things", but I think instead we should say something like "tolerant of all things that do not hurt other people" or some other qualification.

I think that there is a structure here similar to the fact that two negatives make a positive. If I am "not not hungry" then I am hungry. If we add up the "nots" we find that 1 "not" plus another 1 "not" makes zero "nots".

-----

[1] In Russell's paradox, what results is that the supposed "set S of all sets that are not members of themselves" is not an ordinary set, but what we might call a meta-set. Then our two statements:

- If A is a member of the set A, then A is not a member of the set S.
- If A is not a member of set A, then A is a member of the set S.

only apply to normal sets A, not meta-sets. Since S is a meta-set, we now can't just see what happens if A is S, because S is not a valid example of a set A. This avoids the logical collapse.

+	0	1
0	0	1
1	1	0

This is what I call a Battenberg cake structure, as it resembles a Battenberg cake:[2]

This is a mathematical structure that pops up all over the place. It happens if we think about addition of odd and even numbers:

+	even	odd
even	even	odd
odd	odd	even

or multiplication of positive and negative numbers:

×	positive	negative
positive	positive	negative
negative	negative	positive

---

[2] Readers of my previous books will recognize this image. I do love Battenberg.

I think it also comes up if we think about tolerance and intolerance:

- if you're tolerant of tolerance then that is tolerance
- if you're intolerant of tolerance that is intolerance
- if you're tolerant of intolerance that is intolerance
- if you're intolerant of intolerance that is tolerance.

This fits in a Battenberg grid like this:

×	tolerant	intolerant
tolerant	tolerant	intolerant
intolerant	intolerant	tolerant

For me this means that I feel no pressure to be tolerant of hateful, prejudiced, bigoted or downright harmful people, and moreover, I feel an imperative to stand up to them and let them know that such behavior is unacceptable.

A further way to resolve this paradox comes from mimicking the way that mathematicians resolve Russell's paradox using different levels. There, the levels consist of:

1. Collections of objects, carefully defined. These are called sets.
2. Collections of sets; these are sometimes called large sets.
3. Collections of large sets, which we might call super-large sets.
4. Collections of super-large sets, which we might call super-super-large sets.
5. ... and so on.

We could do this with tolerance as well. We could set up levels like this:

1. Things.
2. Ideas about things.
3. Ideas about ideas about things; we might call these meta-ideas.
4. Ideas about meta-ideas, which we might call meta-meta-ideas.
5. . . . and so on.

In this case we could decide that we are going to be tolerant of people's ideas, but not necessarily of their meta-ideas. Their intolerance of other people's ideas would then count as a meta-idea, and we wouldn't feel required to tolerate it.

It is important to be aware that separating concepts into levels can also be used against us, for example in the case of shared knowledge. We could set up levels like this:

1. Things.
2. Knowledge about things.
3. Knowledge about knowledge about things; we might call this meta-knowledge.
4. Knowledge about meta-knowledge, which we might call meta-meta-knowledge
5. . . . and so on.

This arises when allegations of sexual harassment emerge, especially against a well-known figure. Unfortunately it sometimes happens that people in the relevant industry declare that "everyone knew" about this for years. But did everyone know that everyone knew? This would be at the level of meta-knowledge. Sometimes it takes meta-meta-knowledge before victims can unite enough to bring down the perpetrator. This is one of the reasons the aggressors try to prevent communication between victims, with threats and abuses of power, or even a settlement and non-disclosure clause, or other forms of payment. Shared knowledge and meta-knowledge at all levels is an important tool against this sort of manipulation.

It may come as a surprise that thinking about logic and mathematical paradoxes can take us to a discussion of things so apparently distant from mathematics as tolerance and open-mindedness. But to me this is just part of the fact that logical thinking helps us in all aspects of life, even in our personal interactions with illogical humans.

10

WHERE LOGIC CAN'T HELP US

EMERGENCIES, IGNORANCE AND TRUST

CARDIAC SURGEON STEPHEN WESTABY writes in *Fragile Lives* about the fact that if the heart stops, the brain and nervous system will be damaged in less than five minutes. So he often had five minutes or less to decide how to perform surgery. This might not be long enough to do a full logical analysis – only the simplest of logical arguments could be constructed in that time. And there would be no point doing a longer logical analysis if the patient was brain dead when you came to your logical conclusion.

In this chapter I'll start talking about situations in which logic can't entirely help us. We have already seen that logic has to start with something, and that the starting point cannot itself come from logic. But also, we will see that logic also might end somewhere, like a machine running out of fuel. Where logic is concerned, the fuel is usually information. If we don't have enough information to feed into our logic machine, we won't be able to get any further. This can be because of lack of resources, or lack of time, or simply because we are dealing with other human beings and we can't know how they will respond and react.

This doesn't mean we should directly go *against* logic, but it does mean there is a limit to how much we can rely on logic within the given constraints. We will have to invoke something not entirely logical to help us beyond that.

Emotions, gut feeling or intuition can help us make a crucial final leap; that will be the subject of the third part of this book. It's important to understand how far logic gets us and where emotions have to help, rather than pretend that logic can get us all the way there. But we'll start by thinking about where logic can *start* to kick in, which is only after some help with finding starting points. We'll begin with something that's a very natural part of our everyday lives: language.

LANGUAGE

Language has a certain quantity of more or less logical rules. One source of frustration when learning new languages is that there seem to be huge quantities of rules to remember, and also huge quantities of exceptions. It's a tricky combination of logic and non-logic. Some languages are more logical than others. I always enjoyed the logical structure of Latin, but there are still certain things you simply have to remember, such as how the verbs conjugate. At least in English there is little to remember on that front, and no genders to remember for nouns, but there's the awfully confounding question of pronunciation, which is not logical at all. Spanish pronunciation is much more logical (that is, consistent) but still has plenty of exceptions to grammar rules.

We can trace back the etymology of the language we speak now, to see how it came to be the way it is over time, through gradual morphing, borrowing from other languages, and sometimes misunderstanding. But as with logic, at a certain point we get back to a starting point that we can't explain. Many English words come from old German or Latin, but where did those words come from? An etymology dictionary tells me that "cat" comes from Latin but might ultimately go back to an Afro-Asiatic language. Why, at some point in history, did people decide that "cat" was a good way to refer to a small, sleek furry four-legged creature? Some words are more obvious than others, like "cuckoo" which sounds more or less like the sound the bird makes. "Cat" in Cantonese is a high-pitched "mow" (rhyming with cow), which does sound quite like the sound a cat makes. More than "cat" does, anyway. These are the starting points of language and they must once have come from some sort of free or random association. After, all not all concepts make sounds that we can imitate in our naming of them.

One of the difficult aspects of learning a new language is the sheer quantity of vocabulary you have to learn to get yourself started. A certain amount of memorization is hard to avoid. However, as with learning times tables, I have found that memorization doesn't help with actually using the language, because when you're speaking you don't have time to run through your verb conjugation to pick the right one. You have to be able to access it faster than that, from some non-logical, deep-rooted place in your consciousness. We don't learn to speak our native language logically, we do it by immersion, by copying, by emotional connections, and by desire. Children first learn to say things they have a strong desire to say, like "Mama", "Dada", "cat", "ball", "more" or "mine". They often try to proceed logically with language and have to learn that English, alas, doesn't work like that. They might start to notice a pattern with how the past tense is formed, but then they'll say things like "Mummy gived me ice cream".

In any case, children often learn words by adults saying them repeatedly when pointing to something or giving that thing to them. They hear "milk" repeatedly when being given milk, and eventually make a connection. There is no explanation for why that sound goes with that concept: it's a starting point.

FLASHES OF INSPIRATION

Starting points in a creative process might be thought of as flashes of inspiration. You might argue about whether they really exist or not, but I have definitely had moments that I would describe like that. Perhaps it would be less melodramatic to call them an "idea". Where do ideas come from?

Art and music are perhaps the most obvious places where this might happen. I am neither a prolific composer nor a prolific artist, but I have written various pieces of music in my life (some of which I quite like) and made some art of which I'm genuinely proud. In each case, some ideas have just sort of occurred to me. I have no idea where they came from. Some of the music is in the form of songs that I wrote after I read a poem and music just floated into my brain with the poem. That is not logical. There are "logical" ways to develop music, and great composers have many of these techniques at their finger-tips. It can be to do with thematic development, harmonic structure, polyphony where different "voices" are introduced with themes wrapping around one another. Some composers, famously Bach and Schoenberg, used symmetry to transform parts of their composition into new but related music. Unfortunately as a composer I am adept at none of these techniques, so I can only wait for music to float into my brain. This probably explains why I am not very prolific, and why all the pieces I write are rather short.

The rules Bach followed when he was writing harmony were very strict, but that still left him plenty of artistic choices to make within the constraints of those rules. Similarly the rules in sport still leave infinitely many possible outcomes within the rules. The structural rules for Shakespeare's sonnets are quite restrictive but there is still a huge amount of scope for choices and expression while following those rules. The rules narrow down what is allowed, but the rules alone do not determine how the sonnet will go.

Math is another area where flashes of inspiration often start us off. As we mentioned in Chapter 8 on truth and humans, it is an aspect of the non-logical processes involved with thinking up a mathematical proof. Once we have had the ideas we *proceed* using logic, but that part comes afterwards, when we test and exhibit the robustness of our idea. In the next part of this book we will see that this is a valid way to find logical arguments in life as well – start with our instinctive feeling or opinion about a situation, and then try to uncover the logic inside it. It is certainly much more robust than simply declaring all opinions to be "facts".

WHERE LOGIC ENDS

So much for where logic begins. What about where logic ends? Even when we have understood or decided on our logical starting points, or axioms, there might be situations where they don't fully determine what decisions we should make. Imagine choosing from the following menu at a restaurant.

Marinated Roast Breast of Chicken – $18.50

with carmelized pineapple, coconut sauce and wild rice

Pan Fried Ostrich – $21.00

on a bed of ratatouille with a rosemary and redcurrant jus

Char Grilled Fillet Steak – $26.95

with stilton and red wine sauce

Smoked Haddock Fish Cake – $16.95

with spinach, poached egg, sorrel sauce and fries

Roasted Summer Vegetable Tart – $17.95

with truffle oil dressing

All main dishes are served with a medley of fresh seasonal vegetables

---

Perhaps you've decided that you can't spend more than $20, and also you don't like fish. This logically narrows down your options to the chicken and the vegetable tart, but beyond that, logic can't tell you anything. It would be actively illogical to choose the ostrich at this point, but would it be fully logical to choose the chicken? I would call it logically plausible rather than fully logical.

One of the reasons that making decisions is hard is that in most cases logic narrows down the possibilities for us, but several plausibly logical choices still remain. Life is very complicated and much of it is unknowable, with the result that we often end up in situations that logic can't completely decide for us, and it's easy to get stuck in indecision.

To make the decision we can do several things. We can try to add more axioms to the system so that the logical choices are narrowed down to one. For example, we could decide at the last minute that in the absence of any other preference you'd like to try something you've never had, and that means carmelized pineapple. Or you might decide to have the cheapest thing that fits your needs, which would be the vegetable tart. Or you might decide to somehow think about which one sounds the most appealing, by seeing which one makes your mouth water when you think about it. Or you could toss a coin.

Sometimes I'm unable to decide until the very last second, when the server has taken everyone else's order and if I wait any longer I'll just be causing an obstruction. A time pressure is one of the things that helps or requires us to override logic, because logic is too slow.

EMERGENCIES

In an emergency we have to make a rapid decision one way or another. There will be no point making a more logical one if we are flattened by an oncoming truck before we can finish doing the logical deductions. This doesn't mean you should do something that goes *against* logic.

If there is a fire you will, I hope, have an instinctive reaction

"I must get out!" If this is an instantaneous instinctive reaction, it probably wasn't exactly processed logically. But it isn't illogical either. We could at a stretch express this as a series of logical deductions from

A: There is a fire.

to

X: I must get out.

It might go like this:

A is true   (there is a fire).

A implies X (If there is a fire I must get out)

Therefore X (I must get out) by modus ponens.

Unfortunate stereotypes about mathematicians aside, I can't really imagine anyone being pedantic enough to run from a fire shouting "Modus Ponens!"

I can, however, imagine explaining to a child why it is important to escape from a fire. Maybe the child doesn't understand fire yet, so you might have to insert another level of explanation:

Let A = There is a fire.

Let B = I stay here.

Let C = I burn.

Then we have:

A is true.

A and B implies C.

C implies bad.

Therefore I must make sure B is false, i.e. I must get out.

Expressing that in logical terms is a bit over the top but it does show in what way escaping a fire is logical deep down, even if you don't go through those logical steps each time to come to the conclusion, because you have internalized them.

On the other hand I think we can all agree that this would not be a logical deduction:

> There is a fire.
> I will stay here.

Sometimes something starts logical and then by repetition we embed it somewhere deeper in our consciousness so that we can access it more quickly than by logical thought processes. (This is a bit like embedding foreign language verbs into our consciousness by using them, rather than just learning to conjugate them using the rules.) I would say that the conclusion is logical even though we haven't accessed it by entirely logical means.

There seems to be a process by which something logical gets embedded so deeply in our feelings that we then access it by feelings rather than by logic, but if we really needed to explain the logic of it we should be able to turn it back into a logical explanation. Accessing things by feelings is often faster than accessing them by logic, which is why I think one way to become powerfully logical is to convert logic into feelings, just like when you can find your way around a city just by feelings or instinct, without necessarily being able to draw a map or give someone else directions.

INSUFFICIENT INFORMATION

With the menu and with emergencies, logic can run out because there is not enough information around. With menus I often decide I'd like to eat the dish with the least calories, but I can only do that logically if the information is available. Otherwise I have to guess the information and *then* apply my logic.

In emergencies there might be insufficient time to make all the necessary logical deductions or to gather all the necessary information. This can happen in an emergency but it can also happen in sport, where the trajectory of a ball is in principle entirely governed by physics, but we can't take all the necessary measurements in time to do the calculation before needing to hit the ball. It can also be because of lack of resources or physical feasibility. Chess is in principle a very logical game but it is massively complicated by the sheer number of different possible combinations of moves. As a result it is physically impossible to work through all the logical possibilities.

The weather forecast is also, in principle, entirely governed by some laws of physics. But we simply can't gather all of that data. Also there might be so much data and so many tiny variables interacting that the system has in some places become "chaotic". This is a mathematical term that means a system is entirely determined by all the data *in theory*, but in practice it is so sensitive to tiny fluctuations that in terms of forecasts it might as well be random, as we will never be able to collect data accurately enough to avoid those fluctuations. We should not blame the weather forecasters too much when the forecast is wrong, as the system is somewhat beyond the reaches of logic *in practice*.

Another example where we are hindered by complexity and insufficient data is in economics. Economic theories can be hampered by the fact that we don't exactly know how humans are going to respond to certain situations. For example, some people say with certainty that raising the highest level of income tax will not bring in any more income, because the wealthiest people will simply leave the country. This might be true, but we can't know for sure what people would actually do in a hypothetical situation. Anyone who claims to know has, at best, unreasonable certainty about their guesses.

It is possible that we could understand the world entirely logically in principle, but that is never going to happen in practice because we will almost certainly never have enough information. Any consequences involving human reactions to things are almost certain to be guesses about human behavior, rather than logical conclusions.

This is one of the reasons that voting is so complicated in a "first-past-the-post" voting system. This is where the person with the most votes wins, and no second and third choices are taken into account, as in general elections in the UK and presidential elections in the US.[1] If your main aim is to prevent a certain person being elected, you have to guess how everyone else is going to vote, in order to know who is the most likely candidate to beat the one you really object to. The trouble is that if many other people are also trying to guess, the situation becomes rather confused. Deciding in advance who to vote for as a group, in order to mount a robust opposition to a particular candidate, is also tricky. How do you know everyone will act as agreed? Issues of trust cannot be settled by logic alone.

TRUST AND THE PRISONER'S DILEMMA

As we can't have full information about how another human being is going to act, we often have to guess. This is often a question of trust. Do we guess thinking the best of someone or the worst?

You can decide based on past experience, or evidence of their past behavior, but at some level trusting someone else is a leap of faith. Someone who has appeared trustworthy before can always fail later. Sometimes you just have to decide instinctively whether you trust someone or not.

The prisoner's dilemma is a conundrum that examines issues of logic and trust. We imagine two prisoners who have been arrested for committing a crime together, but are locked up separately so that they can be manipulated, and can't confer with each other.

---

[1] In US presidential elections first-past-the-post is currently used to appoint electors of the Electoral College in most but not all states.

Let's call them Alex and Sam again. The prosecutor offers them each a plea bargain. The prosecutor admits that they don't have enough evidence to convict them of the main charge, so in the absence of more evidence they can only convict them of a lesser charge and they'll get a year each. However, if Alex testifies against Sam, they can convict Sam of the worse charge and Sam will get ten years, and in return Alex will be let off. If Sam testifies against Alex, then Alex will get ten years and Sam will be let off. If they both testify against each other then they'll each get five years. If one of them testifies and the other doesn't, they can't suddenly change their mind and decide to take the plea bargain after all.

This is a bit confusing so here is a grid showing the possible actions and resulting outcomes. Sam's outcomes are in the bottom left half of each square and Alex's outcomes in the top right:

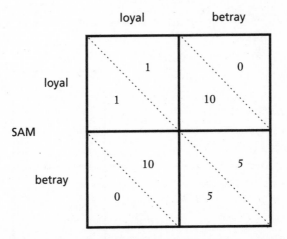

Now, if they both stay loyal they'll both just get one year. But imagine you're Alex thinking about that possibility. If you stay quiet you have to trust that Sam will also stay quiet. What if Sam actually lands you in it and goes free? Then it would be safer for you to testify as well, to protect against that possibility. Meanwhile Sam is thinking the same thing: it is safer to testify, in case Alex can't be trusted to stay quiet. So they both testify, and they both get five years. Whereas if they had both stayed quiet, they would both have got one year. But that option requires trust. You have to trust that the other person is perfectly logical, and also that they trust that you are perfectly logical, and that you trust that they are perfectly logical, and so on. No wonder it's a bit much to expect of real humans.

It can sometimes help clarify our thinking if we consider more extreme versions of the same basic situation. Imagine that some evil opponent is asking a group of people to betray the group. If you denounce the group you will be rewarded $1000 and everyone else will be fined $1000. If someone else denounces the group you will be fined $1000, but if you also denounced the group this just means your reward will be cancelled out to 0. But if nobody denounces anyone at all, everyone will get a reward of $500.

Here the grid for your rewards looks like this:

	nobody betrays	somebody betrays
loyal	$500	–$1000
betray	$1000	$0

YOU

What would you do? If the only other person involved is your best friend then hopefully you know and trust each other enough to know that you won't denounce each other and you'll go home with $500 each. However, imagine doing this with a group of 100 strangers. How likely do you think it is that *nobody* will denounce the group? I would imagine it rather unlikely, and so I am rather likely to do some denouncing myself, in order not to be out of pocket.

In fact, the logic of game theory says that betrayal is *logically* the best strategy in a precise sense. This is gauged by examining what the possible outcomes are if you betray. We don't know in advance what the other person (or people) will do, so we have to consider each possibility and ask whether it would be better to betray or stay quiet. Looking at the table above, we consider each column in turn and see which of our two possible actions produced the better outcome. We see that in both cases betraying produces a better outcome for us. In the case where nobody betrays, we do better by betraying and getting $1000. In the case where somebody betrays, we do better by betraying and getting $0 rather than a fine.

This shows us that in all scenarios for other people's behavior, you get a better outcome if you betray. In game theory this is called a *dominant strategy* and the logic says that this is the strategy that you should take for the best outcome in either scenario. And yet everyone gets a better outcome if everyone can somehow collaborate and do the opposite of the dominant strategy.

Another example where trust and collaboration come into play with rather variable results is the question of climate change. The idea with climate agreements is that all countries cooperate. There is some cost involved with cooperating, but the benefit is global. If nobody cooperates then the effect on the world could be drastic. However, if one country defects and refuses to cooperate, then that country benefits the most – not only do they not incur the costs of cooperating on emissions, but they reap the global benefits of the fact that the entire rest of the world is improving the climate situation. Now, according to the logic of the prisoner's dilemma, we should expect everyone to defect. It is perhaps heartening that this isn't universally the case.

One difference between this situation and the prisoner's dilemma, unfortunately, is the extent to which the different parties actually believe the reward exists. In the case of climate change, some people don't believe there is anything to be gained by cutting emissions, because they don't believe the huge quantity of evidence pointing to the fact that humans are contributing to dangerous climate change. Even if they do believe in it, they might correctly discern that since all the other countries on earth have announced their intention to cut emissions, it might not make that much difference, on a global scale, if the last country does or not. Thus they can save money by not having to make those changes to infrastructure, but still reap the full reward that comes from everyone else having made those changes. This is related to the "commons dilemma", in which a common resource can be used in moderation by a group, but if one person acts selfishly and overuses it they can deplete it, causing an eventual detriment to the whole group including themselves. The commons dilemma focuses more on ongoing situations and different timescales of benefit, whereas the prisoner's dilemma focuses on the curious combination of logic and suspicion causing a system to collapse.

Perhaps the level of trust inside a relationship or a community can be gauged by the extent to which they would be able to cooperate when faced with a prisoner's dilemma. It is interesting that a community's trust and cohesion appears to mean that they can go against the logic of game theory. But the larger the group is, the more fragile that trust is. The cohesion and fragility can happen, at least in principle, at the level of personal relationships, families, communities, countries and the world. I think what this is actually saying is that if a community is infused with enough trust to act as a coherent whole rather than as a collection of selfish individuals, then the logic of the situation changes, and becomes one that can benefit everyone rather than everyone suffering as a result of a few selfish individuals. It shows that there is a sense in which behaving illogically can result in better outcomes than behaving logically. There are situations in which trusting logic alone is not enough, and we would benefit as both individuals and as a group if we trusted in more human aspects of thought as well.

In the last part of this book we will examine what rational humans should do when we are beyond the reaches of logic. We have seen that logic cannot explain and decide everything in the world, so we are going to have to do something when it runs out. We should not pretend that those non-logical things are logical, but we should also not assume that those non-logical things are bad.

PART III

BEYOND LOGIC

11

AXIOMS

LOGIC BY ITSELF HAS no starting point. It consists of a way of making deductions from things we already know. So we have to start somewhere in order to deduce anything logically. Limits are often thought of as being at the end, but there are also limits at the beginning.

We have already touched on this limit to logic that comes from wondering what the root of all truth is. Just like thinking up words for a new language, we have to decide to start with some truths before we can apply logic to find more truths. In mathematics, the things we decide to start with are called axioms, and in life these are our core beliefs.

Axioms are the basic rules in the system. We do not try to prove axioms, we just accept or choose them as basic truths that generate other truths. There are broadly two different ways to approach the use of axioms in math, which I think of as being externally motivated and internally motivated.

The internal approach to thinking about axioms is where we pick some axioms and see what system this generates, logically. In this case, any axioms are valid because we are *assuming* them to be true in the system, to see what will result. The only problem is that if the axioms cause a contradiction then the whole system will collapse and become the null system, in which everything is both true and false. This isn't mathematically incorrect, it just means that there is no sensible notion of truth in that system, so it's not a very illuminating place in which to understand or model anything.

In the real world this gives us a way of conducting thought experiments. We might, for example, imagine a fantasy world in which there is no gender pay gap. Or a fantasy world in which perpetrators of sexual harassment are not tolerated, particularly not in positions of power and influence. It is informative to imagine a world in which everyone reporting sexual harassment were automatically believed. What would happen? First it would mean that huge quantities of women would report the sexual harassment that men inflict on them every day. (Yes, some men are victims too and some women are harassers.) Huge numbers of men would be removed from positions of power. They would either be replaced by women, or by men who would in turn be at great risk of being accused. Perhaps men would start to become very frightened of false accusations. Perhaps employers would start worrying about hiring men in case they were then accused of sexual harassment. If you think that this is an indefensible state to arrive at, it is worth turning it round and remembering that most if not all women fear sexual harassment all the time; we would be replacing this with men fearing *accusation* of harassment. It is also worth remembering that some people are loath to employ women "in case they get pregnant" (even though such discrimination is against the law in many countries); we would be replacing this with companies being loath to employ men in case they commit sexual harassment. We will later discuss using analogies to pivot between different points of view in this way. Imagining a world with these new basic axioms doesn't mean we think it should happen, but it helps us understand the complex web of issues involved. We can understand some things about how far we are from that world and what might need to change in order to get there, together with what some unintended consequences might be.

The external approach is to start with a world that you are trying to understand, say numbers or shapes or relationships or surfaces, or the world we actually do live in. Axiomatizing it is the process of looking for basic truths that logically generate everything else that is true. One famous axiomatization is Euclid's axiomatization of geometry, where he came up with five rules from which all other rules of geometry should be deducible.

The difference between the external and internal approaches is a bit like the difference between moving to another existing country and trying to understand the laws there, as opposed to setting up a new country from scratch and deciding what basic rules you would start with.

There are usually different ways of axiomatizing the same system, so we should consider what a good set of axioms is. First of all, the axioms should definitely be true. They should also somehow be basic, so they should be things you can't really break down into smaller parts. It is often desirable to have as few axioms as possible, but it is also desirable for the axioms to be illuminating, highlighting some important aspect of the structure. Sometimes it is possible to narrow things down to a smaller set of axioms, but at the cost of some clarity.

Usually we don't want axioms that are redundant – if we can deduce one of the axioms from the other ones, then we probably don't need it as an axiom. For some years mathematicians suspected that Euclid's fifth axiom (about parallel lines) was redundant, and tried to prove it from the other four. But they turned out to be wrong – dropping the fifth axiom leaves you with a perfectly good mathematical system, just a somewhat different type of geometry.

Axioms in mathematics are analogous to our personal core beliefs.

WHERE DO OUR PERSONAL AXIOMS COME FROM?

Axiomatizing our belief system in our own lives is somewhat like axiomatizing a mathematical system from an external point of view. We can start by thinking about all the things we believe to be true, and then we can try to boil them down to some basic beliefs from which everything else follows. Anything is valid really, as long as it doesn't cause a logical contradiction, in which case your system of beliefs will collapse. Of course, this is only the case if you are trying to be a logical person. If you are not trying to be a logical person you might be perfectly happy believing a contradiction. But even two people who are logical might disagree about things just because they have different core beliefs – they are using different axioms. It doesn't necessarily mean that one of them is being illogical.

There are then two slightly separate questions: how can we work out what our personal axioms are, and where did we get those axioms from?

We can work out what our fundamental beliefs are by starting with anything we believe and asking why we believe it. The process of repeatedly asking "Why?" is a way of uncovering the deep logic behind something. It is one way of understanding what mathematics is: if we ask why aspects of the physical world work the way they do, the questions may be answered by science. If we ask why science works the way it does, the questions are answered by mathematics. If on the other hand we ask why aspects of the human world work the way they do, we are likely to end up in psychology and ultimately philosophy.

Being able to answer our own "Why?" questions about beliefs requires us to have a certain amount of logical proficiency, so that we can uncover long chains of logical implications, as well as self-awareness. On the other hand, finding out what someone else's axioms are requires us to have logical proficiency coupled with empathy. So we see that an interplay between logic and something more emotional comes in.

My personal axioms fall into three main groups:

1. Kindness: I believe in being kind to people. From this I deduce other beliefs about helping others, contributing to society, education, equality and fairness.

2. Knowledge: I believe in the frameworks that we have set up to access knowledge in various different disciplines. So I

believe in scientific research and historical research, for example, within the confidence levels that those disciplines have established.

3. Existence: I believe we exist, mainly as a pragmatic approach to getting on with life. I'm not so sure about this one and I suspect that if I believed the opposite it wouldn't make much difference, so I've chosen to include it as it seems more helpful than the opposite.

The second point is important to me because it means that I will take some things on trust if I judge that the evidence makes that reasonable. This is not exactly logical overall, because I have not traced all of those conclusions back to their fully logical beginnings. But adding this second axiom means that I have traced those conclusions back to their fully logical beginnings inside my axiom system. For example, I believe in gravity although I don't understand it logically. So I don't know how to trace it back to first principles in mathematics, but I do know how to trace it back to the axioms of my personal belief system as it comes from my belief that scientists are probably right about it.

I think we have to accept some starting points in our logical system in order to get anywhere. This is true in mathematics and life. The important thing is to be clear what they are. As we'll see, this can help us identify more complicated beliefs, and also pin down why someone else might disagree with us about some more complicated beliefs.

WHERE DID WE GET OUR AXIOMS?

The question of where we came up with our own axioms is more philosophical, but important as it may help us understand other people who disagree with us fundamentally. Most of us get our personal beliefs from some combination of our upbringing, society, education, life experience and gut feeling. Some things are instilled in us by our parents, but most of us don't have exactly the same opinions as our parents, which means something else must influence us. Education can expand people's world view and lead them to see things differently from their parents. So can life experience, especially if the parents grew up in a very different era, culture, or economic environment. Some beliefs seem to come from nowhere in particular except personal conviction, but if we think about it very hard, we might be able to see where that personal conviction comes from.

For example, I believe that it is more important to be kind than to be right, and I simply feel this very strongly. But if I think about it hard, I see that it comes from my life experience, and incidents when I've been so hurt by other people's behavior that I've come to believe more and more strongly in the importance of kindness.

I believe that education is the most important way in which I can contribute to the world, and again, I simply feel this very strongly. But if I examine the source of that feeling, it is a combination of values instilled in me by my parents and piano teacher, together with evidence that I am not really cut out to be a doctor (too much memorizing in medical school) or to brave war zones to rescue people (I am too afraid of physical danger).

Some people can trace all their fundamental beliefs back to religion. That still leaves the question of where their belief in religion comes from. For some people it might come from their parents and their upbringing, possibly reinforced by their education. For others it comes from a particular time in their life, unfortunately often after a tragedy or trauma. It might also come from being swayed by a very influential person. Understanding what people's fundamental beliefs are helps us find the root of disagreements, and understanding where they get their beliefs can help us understand how we might be able to address those beliefs.

MORE AND LESS FUNDAMENTAL BELIEFS

One way to axiomatize a system of beliefs is to take every single belief as an axiom. This certainly means all your beliefs can be derived from the axioms using logic, but it has not achieved anything. It would be a bit like a recipe for lasagne where the only ingredient is "lasagne". Rather, the point of axiomatization is to understand the roots of a system and what holds it together.

Taking all beliefs as axioms would obliterate the need to follow any sort of logical deduction, and although it's not exactly illogical (it doesn't contradict logic) it's hardly a well-developed point of view. It would be a rather extreme situation, but we do meet people who are unable to justify certain beliefs of theirs very much at all – they take quite complex beliefs as fundamental, without justification. Or in some cases there is some justification but it doesn't get us very far. For example, a person might say, "I oppose same-sex marriage because I believe marriage should be between a man and a woman". This might sound like a justification as it does have the word "because" in it, but it's really just a restatement of the initial belief.

Our fundamental beliefs are rooted in something beyond logic. However, sometimes abstraction can help us find something more fundamental inside what we believe. As I discussed in Chapter 2, I have discovered that my belief in taxation and social services is rooted in a more fundamental belief that false negatives are worse than false positives. We will come back to this in Chapter 13.

As a related issue, I believe that everyone's situation is a combination of their own input and their circumstances. I believe that humans are not isolated creatures, and that we exist inextricably connected to the communities around us, and thus there is some collective responsibility for both success and misfortune. This in turn comes down to my more fundamental belief that we should understand all things in terms of systems rather than individuals, as in Chapter 5, whether this is people, factors causing a situation, or mathematical objects – the latter being why I do mathematical research in category theory, a discipline that focuses on relationships between things, and the systems those form.

Having said that, I do believe that everyone should take responsibility for themselves, while at the same time believing that we should all look after each other. Where exactly personal responsibility ends and collective responsibility kicks in, I am not quite sure. This does however lead me to another more fundamental belief: that there are many gray areas in life and that it's important we understand them for what they are rather than ignore them or force them to be black and white. This might mean that we can't be entirely logical where gray areas are concerned. In the next chapter we're going to look at the different ways that logic has of dealing with gray areas, and see that some of them are rather undesirable, pushing us to extremes in ways that don't feel right. We will see that as we move through a gray area things may start out seeming "the same" and then gradually morph into something that seems different, although somehow the framework was the same. In Chapter 13 we will take this further and analyse how analogies work, and how we use them to pivot between things that aren't the same, by means of some way in which they *are* the same. A big issue with analogies is deciding when they count as a good analogy or when they have been pushed too far. In Chapter 14 we will talk about when things should count as the same or not. False equivalence is a widespread logical fallacy, but in between true logical equivalence and false equivalence is a gray area of things that are the same in some sense, we just have to find it. In Chapter 15 we'll look at how we can use all these techniques to engage our and other people's emotions, to try and better see eye to eye with other humans, and finally in the last chapter we'll paint a portrait of a good rational human – not a perfect computer – and what good rational arguments should look like.

12

FINE LINES AND GRAY AREAS

HOW LOGIC PUSHES US TO BLACK AND WHITE IF WE'RE NOT CAREFUL

ONE NIGHT DURING MY first term at university another fresher was found in the kitchen just before midnight eating a bowl of cereal. He explained that, according to the best-before date, his milk was about to go off at midnight. I'm afraid we laughed at him rather heartily, and in his belief that a best-before date is so precise, or that milk could suddenly go off at midnight like Cinderella's carriage turning into a pumpkin.

Unfortunately many more serious problems arise from our attempts to deal with things that are on a sliding scale. If we're not careful logic pushes us to extremes, so if we want to avoid adopting extreme positions all the time we have to do something more human. A human approach is more nuanced than that. It turns out that our brains are able to process gray areas in subtle ways that don't seem entirely logical but do seem to make sense. Rather than use logic to push out the sense, we should find the logic inside the human nuances. There are various different ways of dealing with gray areas that stem from different logical interpretations. In this chapter we'll discuss these different approaches, and the pitfalls of adopting the simplest approach of drawing a line. Allowing for some uncertainty can seem unsettling, but it can avoid both the extremes and the anomalies of drawing a line.

One of my favorite moments in Jane Austen's *Pride and Prejudice* is when Elizabeth asks Mr. Darcy how and when he fell in love with her. He replies,

I cannot fix on the hour, or the spot, or the look, or the

words, which laid the foundation. It is too long ago. I was in the middle before I knew that I had begun.

He is unable to draw a line and say that is the exact line in between not being in love with her and being in love with her. Where could such a line possibly be? He can only know that at some point he was not in love with her, and at some later point he definitely was. This is understandable because being in love with someone is a somewhat hazy and nebulous concept that grows gradually (except in cases of love at first sight, if that really exists), with a lot of gray area in between the definite "no" and definite "yes".

Gray areas are an important part of the human experience, but are not very well dealt with by logic. Logic is about getting rid of ambiguity. The law of the excluded middle forces us to put the whole gray area in with black or in with white. That is better than the sort of black and white thinking that pretends the whole gray area doesn't exist, but it can have the same effect as it can push people to think only the black or the white if they try to pursue logic to its rigid conclusion.

The language we use in everyday life seems to be getting pushed further and further to extremes and certainties. People say that something is "the best thing ever" (or the worst). They try to reassure me by saying "Everything is going to be fine", they advertise events to me by saying "You will not want to miss this!" Admittedly, "You might not want to miss this" doesn't sound so good. But I worry about the world turning into certainties that are *almost* certainly flawed. We should understand different ways of dealing with gray areas and become better at working with their nuance, instead of longing for the false promise of black and white clarity.

CAKE

I am not immune to extreme thinking myself. Here is the kind of reasoning that makes me prone to getting fat:

- *It won't hurt to eat one small piece of cake.*
- *And however much cake I've already eaten, it can't hurt to eat one more mouthful.*

Unfortunately, logically this means that it is fine to eat *any* amount of cake, as long as you go one mouthful at a time. And unfortunately this is what I am prone to do.

It is another example where being strictly logical is not entirely helpful. The only way to avoid thinking it's okay to eat infinite cake is to decide it's not even okay to eat one small piece of cake. I am much better at avoiding cake altogether than eating a small amount and stopping. The trouble is that everyone around me is likely to point out that one small piece by itself is fine.

This is an example where the logic of the situation pushes us to one of two extreme positions, either

- it is not okay to eat any cake at all, or
- it is okay to eat infinite amounts of cake.

The trouble is the gray area. There is no strict line we can draw between a sensible amount of cake and "too much". Parents are liable to try and draw those lines to stop children gorging on cake, but children aren't fooled – they easily see that those lines are arbitrary, and they try to push them by having one more mouthful, and one more. Or they try to push bedtime by an extra two minutes, and then another extra two minutes, by saying they need to go to the toilet, or fetch a toy, or drink some water, or any number of other spurious requests. But really, the bedtime itself is spurious – it is an arbitrary line that has been set in a gray area between "sensible bed time" and "much too late".

One way to get round this logic is just to shrug and say just because something is logically implied, that doesn't mean I'm going to believe it. But that is unsatisfactory as it would allow for all sorts of other illogical thinking, such as believing two things that cause a contradiction.

The idea of believing all the logical implications of your other beliefs is called "deductive closure" – a set of statements is deductively closed if it also contains everything you can deduce from all the statements in the set. So my set of beliefs is only deductively closed if I believe all the implications of my beliefs. I think this is an important part of being a logical human being, and I'll come back to it in the last chapter.

So, if I want to be rational, what can I do about gray areas? I might have to let go of the most obvious logical approach and learn to deal with something more complex.

DRAWING A LINE

Gray areas are often dealt with like bedtime – an arbitrary line is drawn somewhere in the gray area and a rule is made. Where you put the line in the gray area depends on how dire the consequences of the extremes are. If one of the extremes is very dire then the line in the gray area will probably need to be further away from that extreme, to give a buffer zone around it. For example, some roller coasters have a minimum height requirement for safety reasons. If you're too small then the safety harness won't fit around you enough to keep you safe. In that case the consequences are very dire (injury or even death) and so the limit should probably be at the tall end of the gray area, to be safe.

With the cake, I am trying not to get fat, so I should probably draw the line safely within the "most probably won't make me fat" part of the gray area, rather than somewhere in the "might not make me fat but it's not clear" part. This is especially true because I am liable to stray over my line a bit, so I should put a little buffer zone in to be on the safe side.

One contentious area of drawing lines I've had to deal with a lot professionally is grade boundaries in exams. In the UK system, students graduate from university with a degree that is classed as first class, "upper second" class, "lower second" class, or third class, otherwise known as first, 2:1, 2:2 or third. But where should the boundaries be drawn? I have spent long and contentious hours in examiners' meetings battling this out, in what is essentially a futile exercise placing a line in a gray area. No matter where you place it, someone will argue that it's unfair to the person just below it, and as a result the line tends to shift further and further down. There is no logical place to put that line. I think the only logical thing to do is get rid of the lines and publish either averages on a fully sliding scale or percentiles instead.

A more contentious gray area arises when talking about race. I will not call this "literal" black and white, because we're actually all shades of brown and pink. As discussed in Chapter 4, Barack Obama is often called "black" although one of his parents was black and one was white. So arguably it's just as valid to call him white. However, once we understand that "black" here is being used to mean "non-white", we see why it makes some sense to call Obama the first black president of the United States. Which is to say, we have found *the sense in which* it makes sense.

Where should we draw a line between black and white? If we draw it nearer the white side, this could acknowledge that only people who look really white enjoy the privileges of white people. But it could constitute exclusion of anyone non-white as "other". At least talking about white people and non-white people is a genuine dichotomy, where black and white is a false one.

SEXUAL HARASSMENT

It can be especially hard to draw lines when dealing with people who do not respect boundaries. This can happen if you are the kind of person who likes being generous and being able to help people. Unfortunately people are liable to ask more and more of you. More seriously, it arises in situations of micro-aggressions and sexual harassment: how serious does someone's misconduct need to be before you should take action and report them?

Some forms of physical contact are generally accepted, such as a handshake. Others are clearly not, such as groping. But where do we draw the line in between? Is touching someone's shoulder appropriate? Their back? Their waist? Their hip? Do we have to draw a literal line on our bodies to indicate where can count as friendly and where counts as harassment? This is a difficult dilemma for those who have been made uncomfortable by someone's behavior, especially if they're in a position of vulnerability or on the lower side of a power differential. If you report someone touching your shoulder you will almost certainly be told you are overreacting. So at what point is it worth taking action? If you accept one action then you can feel like the next small escalation isn't that much worse. But all those small escalations add up.

Manipulative people can actually exploit this to take advantage of people who are prone to being kind, generous or accepting. As soon as you give in once, even a little bit, they know that the boundary can move by a little bit, and so by applying that logic repeatedly, they can think that the boundary is totally movable. If you put your foot down and stop them at some point, they might accuse you of being mean, unreasonable, overreacting.

Mr. Darcy's inability to draw the line for love also applies to hurt – we don't necessarily know where is the exact moment that something is going to start hurting us. We only know where it will definitely hurt us a lot, like if inappropriate physical contact becomes rape.

It took me a while to learn that the best way to be safe is to draw the line somewhere where you definitely feel safe, before things start being at all questionable. This creates a buffer zone encompassing the gray area, protecting us from the area that is definitely dangerous, but without definitely declaring where the buffer zone ends and the unsafe zone starts:

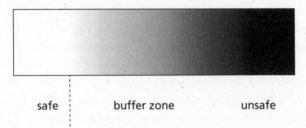

safe   buffer zone   unsafe

I used to think this was ungenerous, because it means that wherever I draw the line there might have been a little bit more I could allow without getting hurt. But having been hurt too many times I now know I need that buffer zone, like with the cake, to protect myself. And I also know that protecting yourself is important, and not necessarily ungenerous. As we saw in Chapter 5, if protecting yourself means denying someone something, or even hurting them, I don't think it's your fault if the other person then gets hurt by it. It's the fault of the system, of the toxic relationship that has created such a zero-sum game.

BODY MASS INDEX

Body Mass Index is a useful but rather flawed measure of healthiness where you take your mass (in kilograms) and divide it by your height (in metres) squared. The first thing that people object to about this is that it doesn't take into account how muscular you are, and so very strong athletes are liable to have a high BMI as muscle is very dense. However, personally I find this argument spurious as it's pretty obvious whether you are a muscular athlete or not. More to the point (and to avoid making a logically fallacious sweeping statement), I know perfectly well that *I* am not a muscular athlete. I don't need to use calipers to know that I have fat on me, even if I hide it well enough under my clothes that people insist that I surely have no fat at all.

The other problem is that arbitrary lines have been drawn for what counts as a "healthy" weight in terms of BMI. The cutoff for women is usually stated as 25. But of course, it's a sliding scale. It doesn't mean that someone with a BMI of 25 is fat and someone with a BMI of 24.9 is fine. It's supposed to be a guideline, and I am quite happy to use it as a guideline. It does create silly situations when the doctor weighs me, though, because sometimes I know that my shoes are likely to tip me over the BMI 25 mark so I insist on taking them off. If they register a BMI over 25, the computer automatically puts an alert on my file, and even though the doctor shrugs and says I'm so close that there's no need to worry, I have made such a big effort to lose weight and keep it off that it is utterly galling to register as overweight even if I know it was really just my shoes.

Despite such escapades I still think it's better than having no guideline and doing what I used to do, which is convince myself that I was still fine really, even though I was gaining 20lb a year. This involves another gray area, to do with reasonable weight gain. You might shrug to yourself and say "Gaining a couple of pounds in a month isn't so bad, not worth worrying about." But then if you say that to yourself every month you'll find that you're gaining 24 lb per year. Personally at some point I imagined myself in 10 years (instead of just thinking month by month) and finally realized that I had to draw a line somewhere, even if it was arbitrary. I try to draw it on the safe side of the BMI line, somewhere around 24 rather than 25.

One interpretation of what I'm doing is treating the line

itself as something hazy, so I try to stay far enough on the good side of the line that I'm out of its hazy range.

INDUCTION

These arguments by small incremental steps are related to the principle of mathematical induction. This is different from argument by induction, which is a flawed type of argument where you generalize from a small sample to a large one. For example, "The sun has risen every day of my life so far, so it will rise tomorrow."

Mathematical induction is logically secure, and is a bit like climbing up steps. Babies learn to climb up one step and then discover, with delight, that if they just repeat it they can go up whole flights of stairs, possibly all the way to the sky. All they need is for someone to put them at the first step (and for nobody to intervene and take them off the stairs).

Mathematical induction says if you know something is true for the number 1, and if you have a way of climbing up by 1, then you know it's true for all whole numbers. If we apply this to small cookies, we'd say

- It's fine to eat 1 cookie.
- If it was fine to eat some number of cookies, it's okay to eat 1 more.

Therefore it's fine to eat any number of cookies.

Mathematical induction is stated in terms of whole numbers n, and we say that we are trying to prove that some property P is true of each number n. So $P(n)$ might be the statement "It is fine to eat n cookies." Then the argument looks like this:

- $P(n)$ is true.
- $P(n) \implies P(n + 1)$.

Then by the principle of mathematican induction, $P(n)$ is true for all whole numbers n.

This is fine for whole numbers, but it gets tricky if you're trying to deal with a sliding scale that includes all the numbers in between, or even just all possible fractions. This is because there is no smallest unit of "jump" for us to take steps in.

We might try to apply this to a series of cookies that all seem to be "more or less the same size". Perhaps this means that they are within 5g of each other's weights. You might be happy thinking that a 50g cookie is more or less the same size as a 52g cookie, and that this is more or less the same as a 54g cookie, but after a few steps like that you will end up with a cookie twice as big. I have done this by handing out cookies to a class of 20 students without telling them what the point is. I ask them all to compare their cookie with the person next to them, and they are all happy that their cookie is more or less the same. But then I get the first and last student to compare cookies and we all collapse in giggles because the first one is tiny and the last one is enormous.

FUZZY LOGIC

I think one way to deal with this is to allow for more nuanced levels of truth. With cookies, the point is that there isn't just "fine" and "not fine" for amounts of cookies to eat. There is "fine", "less fine", "sort of okay but not great", "not that great", "dubious", "suspect", "a bit much", "too much", "way too much", "ludicrously excessive" and so on. As we saw in Chapter 4, normal logic with the law of the excluded middle doesn't allow us anything except "fine" and "not fine", so we end up pushing everything into one or the other as we can't find a logical place to draw a line. Instead we might try to treat truth values as something on a scale between 0 and 1. This can be dangerous in some situations because it can give the idea that truth is something negotiable, and that some things are more true than others. However, I think this is true in the case of gray areas. It also might be true in the case of probability, when we can't be certain what the truth is, we can only be a certain percentage sure, with the rest being in some doubt. Percentage probabilities are, in a way, placing things on a scale of truth between 0 and 1. Unfortunately it often seems that we humans are not very good at understanding those either.

In Chapter 4 we briefly mentioned fuzzy logic, a type of formal logic that takes truth values in a range from 0 to 1. This measures the extent to which something is true, rather than our certainty about whether or not something is true. The two things are related, but not exactly the same. For example, if I look up the weather online it gives me a percentage likelihood of rain during each hour of the day. Usually I conflate this with an amount of rain in my head – if the forecast says 90 percent chance of rain, I interpret it to mean it's probably going to rain hard. If it says 40 percent chance of rain, I interpret it to mean it might rain a bit. In practice this is likely to be because of where the uncertainty of a weather forecast comes from. The only way to be really sure that rain is coming is if there is a very strong storm heading strongly in this direction. If we're not sure, it might be because it's only a light rainstorm that has some chance of fizzling out or changing direction.

Similarly if an exam is marked with just pass and fail, then the result is very clear cut. Before you get the result you might be unsure about whether you passed or not, but only if you're a somewhat borderline case. If you're a really excellent student you might be unsure quite how well you did, but still sure you passed. Again, the certainty and the extent to which something is true are related.

However, even once uncertainty has been eliminated, the extent to which something is true can still vary. If an exam is marked on the whole scale from 0 to 100, you might get a mark of 71 and ask your teacher if that counts as good or not. There is now a whole scale from good to bad, and the uncertainty comes from the gray areas, not from actually not knowing.

Fuzzy logic is currently used more in applied engineering

than in math, to deal with gray areas in the control of digital devices. One example is in rice cookers, where the cooking process can be adjusted according to some slightly vague conditions like whether the water is being absorbed slowly, quite slowly, quite quickly, or quickly. This can also be done for heating or air-conditioning control, or anything else that needs to respond dynamically to potentially changing conditions. Of course, a definition still has to be produced as to what those gradations mean, but having the possibility of truth values in between sheer true and false opens possibilities for more subtle control of devices.

THE INTERMEDIATE VALUE THEOREM

Another way to deal with fine lines and gray areas that isn't fully logically determined is to acknowledge that the line is somewhere in the gray area and we don't know where it is; we just know it's somewhere in that area. We can put bounds on the area by pointing to one place that's definitely below the line and one place that's definitely above the line.

This is how you can make a batch of cookies so that everyone has their perfect size of cookie. You can start with a cookie that is definitely too small, maybe just 2g of cookie dough. Then you can make them get very gradually bigger so that each one seems about the same size as the previous one, but they're getting imperceptibly bigger. Keep going until you reach one that is obviously definitely too big, say twice as big as your face. Opposite is a set of such cookies I made.[1]

Because there is essentially every size of cookie in between, this means everyone's perfect size of cookie must be in there somewhere. This helpfully deals with the fact that my personal favorite size of cookie is smaller than most people's. This way I can meet my own needs and other people's at the same time,

---

[1] Photo credit: M. N. Cheng

without actually having to know what anyone else's perfect size of cookie is – I can be sure it's in there somewhere.

This is an application of the intermediate value theorem, a theorem in rigorous calculus that math students usually study as undergraduates. It says that if you have a continuous function that starts at 0 and goes up to some number *a*, it must take every value in between. What "continuous" means here is rather technical, but it basically means there are no gaps. You might argue that my cookies do not take *every* size, otherwise there would be infinitely many of them. That's true, but I am using a real-life approximation to the intermediate value theorem. Really I'm saying that the whole thing is true up to a certain accuracy determined by our perceptions.

A few months ago I was talking to one of the art students at the School of the Art Institute of Chicago. She was exploring people's perceptions of reality by creating visual illusions and trying to see if viewers would believe they were physical constructions, or think they were digital manipulations. The question was to find the sweet spot where people would be really unsure which it was. I realized that she could invoke the intermediate value theorem: make a series of pieces starting with one that was obviously a physical construction, gradually becoming less obvious until she ended with one that was obviously physically impossible so must be a digital manipulation. At some point in the gray area in between there would have to be a point where viewers were unsure whether it was real or digital. The artist would not need to know exactly where it was, and indeed it could be in a different place for different viewers. All the artist would need to know is that it is somewhere in that gray area.

In a way this is also what chocolate makers do with the percentage cocoa content in their chocolate. They make a whole range of different percentages so that all chocolate lovers will find their perfect percentage in there somewhere. I regret, however, that so many chocolate makers stop their range somewhere around 70 percent (in fact 72 percent seems a popular place to stop), because that does not encompass my own personal favorite spot. My gray area for chocolate is somewhere between 80 percent to 100 percent, depending on my mood. As a result, to get my perfect chocolate, like my perfect size of cookie, I usually have to make my own.

This is a similar principle to the one that is giving us more and more finely tuned gradations of race. In the unenlightened old days there was just "white" and "non-white". We now talk about "people of color" but we also talk about mixed race, although this often implicitly means mixed between white and non-white. People have been coming up with words for mixed races between non-white and some other non-white, such as blasian for black and Asian. A friend of mine calls himself "Mexippino" for Mexican and Philippino.

But should we have different words for someone who is one quarter Asian and three quarters white (who is likely to be quite white-passing), as opposed to three quarters Asian and one quarter white (who might look entirely Asian)? As described throughout this chapter there are various possibilities, each of which has advantages and disadvantages. We can talk about "white people" and "Asian people" which is simple, but results in the exclusion of those who fall into neither category exactly. We can draw a line somewhere and end up creating bizarre anomalies, where someone is classified as a "person of color" although they look white to most people. We can create increasingly many finer and finer categories taking everyone into account, but ending up with an intractable collection of extremely specific descriptors. We can unify everyone, ignore race, and declare that we're "all humans", as some people do when they call themselves "color-blind". But this discounts people's real experiences of racial discrimination. I think that above all we should acknowledge that there are gray areas and become more comfortable with accepting them.

BRIDGING GAPS

All this is to say that if we aren't careful about gray areas and logic we can end up being pushed into taking extreme positions, as the only way of remaining logical. Indeed, it's possible to use incremental steps of logic to argue someone into a very extreme position without them realizing what is going on. If we hold people to this black and white logic then we push all disagreements into further and further extreme opposites. I think, instead, we should give everyone ways to get out of these positions that are still counted as logical. Whether it's fuzzy logic, probabilities, hazy lines, or lines simply placed somewhere unknown in a gray area, or simply being more comfortable with less certainty, these are all more nuanced ways of dealing with our very nuanced world. The gray area is a bridge between black and white. There are few things as simple as black and white in the real world. Really we're all living somewhere on that gray bridge, and for some people it feels unsettling to take a position of nuanced uncertainty. If we all acknowledge that, and even build more bridges, I think we will achieve better understanding.

We have discussed how gray areas can hurt us when exploited by unscrupulous people, but we can turn this around and exploit gray areas to help ourselves, if we apply them judiciously. The idea is that we can use small, unnoticeable increments to nudge ourselves gradually to somewhere that is quite far from where we started. If we tried to do it in one leap it would seem insurmountable. I use this psychological trick all the time to make progress. I use it when I'm learning a new and difficult piece of music at the piano. I start by only being able to play it extremely slowly, but I use a metronome and gradually increase the speed by a tiny and unnoticeable amount each time. I might start it at a rate of 40 beats per minute, at which speed the piece is easy, and then go up to 42. My fingers don't notice any difference. Then I go up to 44, and my fingers still don't notice any difference. It doesn't take that long to double or even triple the speed at which I can play the piece. The principle of "a little bit more won't make a difference" is harmful when it's about eating cake, but beneficial when it's about accomplishing a big task.

More seriously, this method can be used to find a bridge between apparently opposing ideas. We have discussed the difference between believing in social services and not believing in them, in terms of false positives and false negatives. One person believes in helping everyone who needs help, even if that means helping some extra people by mistake. Another person believes that everyone should take responsibility for themselves. This argument can be very divisive, but we could instead acknowledge that there's a gray area. The person who believes in social services probably doesn't believe in simply giving out large amounts of money to everyone who asks for it, without question. The person who believes that everyone should take responsiblity for themselves might be able to acknowledge that some particularly "worthy" people need help, perhaps members of the military who have been injured during active service. If this is so, then we have established that the question is not whether or not we should help people, but to what extent we should, and under what circumstances. It is now a question of where we place people in the gray area and how we treat the gray area. In the coming chapters we will discuss this technique of pushing a principle to extremes abstractly in order to encompass those who seem to disagree with us, and draw them onto the bridge that is the gray area. The first step is to think about understanding a difficult argument by comparing it with a more understandable one that has something in common, that is, it is analogous in some way. This is the subject of the next chapter.

13

ANALOGIES

WE HAVE SEEN THAT abstraction is how we get to the world where logic works. The abstract world is a world of ideas and concepts, removed from our concrete, messy world of objects, humans and feelings. But how then does that logical world interact with the world we actually live in? Understanding logical situations is all very well but one of its limits comes from the fact that the abstract world sheds light on our "real" world, but is not actually our real world, and so something is bound to be lost or distorted when we move back to the real world.

In this chapter we'll talk about how abstract ideas interact with real world situations in the form of analogies, how analogies can help us understand both good and bad arguments about our world, and what some of the pitfalls are when we invoke analogies.

ABSTRACTION

At the beginning of this book we discussed the fact that nothing in the real world actually behaves according to logic. So in order to study anything using logic we have to perform some abstraction, that is, ignore some of the details of the situation so that we move to the abstract world of ideas, where things do work according to logic. This is sometimes like modelling a situation with a simplified version, or like focusing on just certain aspects of a situation. Doing this abstraction enables us to use logic, but the process of abstraction is not itself logical – we have to choose what to focus on and how to simplify. In the previous chapter we saw that there can be many different ways of finding an abstract version of the same situation. This doesn't mean that some are right and some are wrong, it means that different abstractions show us different things, and we should be aware of what we have lost and gained by doing it.

When we forget details about situations, many different situations start looking the same. Abstraction is a way of finding what different situations have in common, just like we did with the cubes and cuboids of privilege in Chapter 6: we discovered aspects of various different situations that could be understood in cuboid-shaped interactions.

This is why abstraction is linked to analogies: an analogy is a similarity between two different situations, and abstraction takes that similarity very seriously and treats it as a situation in its own right. One of the most basic analogies is between objects like two apples, two bananas, and two chairs. The thing that those situations have in common is the concept of "two", which comes from taking what is analogous between those situations and regarding it as a concept in its own right. Arguably all of mathematics comes from finding similarities between different situations in this way, and when we study them as concepts in their own right we have moved one level more abstract than before.

Following what we learned in Chapter 6 about the efficacy of highlighting relationships between things, we could draw these relationships like this:

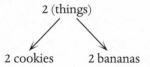

Here the arrows represent the process of going from something general or abstract to an instance of it. Abstraction is a way to pivot between different situations that have something in common. In fact, I've drawn the diagram of abstraction to look a bit like a pivot. The number two enables us to pivot from a situation involving two cookies (perhaps one and then another one) to a situation involving two of something else.

Moving to the abstract level is one way to find the logic in a situation, but making an analogy is a way to do it without explicitly going through the abstract world. Usually in normal life we make the analogy without being explicit about what the abstract version is. This can be helpful in real-world situations where exhibiting abstract logic would be rather pedantic – precise, without illuminating the situation unless you're talking to others trained in abstract logic. In fact, the way we teach small children numbers usually involves showing them sets of two things over and over again and encouraging them to find what is analogous between them themselves.

By contrast in math great power comes from making the abstract version explicit, as we did for the cuboid of privilege. It means we can make analogies that are more complicated or subtle, or enable us to encompass examples that are much further from where we started. We're really just like monkeys jumping from tree to tree and discovering that we can get further if we swing from a branch rather than just jumping. In fact, this is an analogy about analogies, so we will now look at the abstract principle behind analogies themselves.

FRAMEWORK FOR ANALOGIES

Whenever mathematicians feel like they're doing the same thing over and over again, they look for an abstract version that represents the situation. I find myself making analogies repeatedly, so what about the abstract version of an analogy? The general situation is that we are making an analogy between concepts A and B, via an abstract principle X that is often implicit rather than explicit. The diagram looks like this:

As with monkeys swinging from branches, we should think about what level of abstraction we choose to pivot from. The more details we ignore about a situation, the more things become the same. Math tends to get more and more abstract as it goes on, gradually losing people at each stage as they become uncomfortable with the level of abstraction. People often tell me that they lost it when "numbers became letters". This abstract pivot could be shown like this:

Here a and b represent the numbers $1, 2, 3$, or something else. But there's a level even more abstract than that, where we draw an analogy between addition and multiplication themselves, and think about "binary operations", which include addition and multiplication but also many other things. Here is the further level of abstraction, with the symbol $\odot$ representing a binary operation that could be $+$, $\times$ or something else:

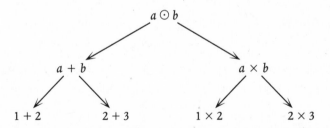

At the first level, $1 + 2$ is analogous to $2 + 3$ but not to 1×2. At the second level addition itself is analogous to multiplication, so everything on the bottom row can now be treated analogously.

For some purposes the top level is a good level, but for other purposes we should only go as far as the middle level.[1]

One of the important lessons my PhD supervisor Martin Hyland taught me was the importance of finding the right level of abstraction for the situation. This is like shining a light at an appropriate distance so that you can see enough detail but also enough context around the thing you're looking at. In a way, for abstraction, this consists of forgetting as many details as possible while still retaining the truth of what you are trying to study. If we forget details that are relevant, we might forget something that is critical to the situation. After all, if we forget enough details eventually everything becomes the same, and that is arguably not a productive way to look at the world. (Although I do think we gain something from remembering that all humans are at root the same.)

However, if we don't go abstract enough, we might miss the chance to draw some links between more things. This happens with numbers in this diagram:

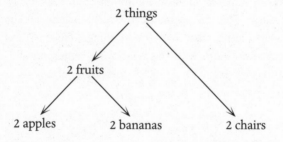

If we only go up to the level of "2 fruits" we will get an analogy between 2 apples and 2 bananas, but will omit 2 chairs. In order

---

[1] One way in which addition and multiplication are not analogous is if we consider inverses. Adding a number can always be inverted (undone) by subtraction. But multiplication by a number can't always be undone – we can't undo multiplication by 0, because we "can't divide by 0". What that really comes down to is that multiplication by 0 results in 0 wherever we start, so if we try to reverse that process we have no way of knowing where to go back to, whereas with addition we always know.

to include "2 chairs" we need to go up further, to the level of "2 things". This is what happened when we found the cube of privilege in Chapter 6. We started with the cube for factors of 30, and the one for factors of 42, and we found that they had the same shape because both numbers are a product $a \times b \times c$ of three different prime numbers. The analogy is expressed in this diagram:

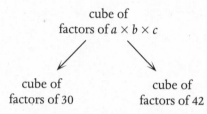

But then we realized that if we go one level more abstract and think of it as a cube of subsets of $\{a, b, c\}$ then the analogy applies to many more things, including any three types of privilege. We have gone up a level of abstraction as here:

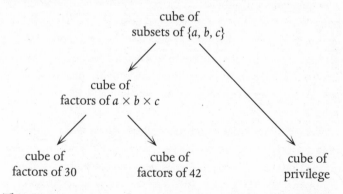

This gives us a way of expressing the possibly surprising fact that although thinking about something *more* abstractly appears to take us *further away* from concrete ideas (vertically in the picture), it enables us to pivot further away from where we started (horizontally), and thus encompass *more* ideas, including more concrete ones. Much of my argument about math comes from my view that math is a little removed from normal life, so if we pick low pivots we will not get very far out of math, perhaps only as far as physics. But if we get more abstract we can pivot far from math and apply our analogies to very real situations in life. Here is an example that is very real in life:

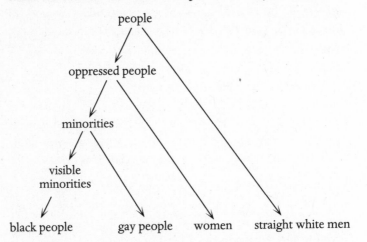

The question is whether the experience of the people along the bottom row is analogous. A nuanced answer is: it depends how far up the levels of abstraction you go. Unfortunately, divisive arguments often arise because everyone picks the level of abstraction that best suits their argument and refuses to consider the possibility that other levels could in any way be valid.

We've already discussed many abstractions and analogies – in fact, perhaps the whole book is about how carefully chosen abstractions and analogies can generally shed light on all our arguments in the world. But here are some specific ways in which analogies can help us.

FINDING AXIOMS

In Chapter 11 we talked about finding axioms for our personal belief systems, that is, the fundamental beliefs from which all our beliefs stem. Thinking about analogies can help us understand what our personal axioms are, or what someone else's personal beliefs are.

In Chapter 2, I discussed discovering my own axiom that I care more about false negatives than false positives. It came from thinking about believing in social services, and the sense in which that is analogous to the following wide variety of situations.

When arguing about affirmative action on grounds of race, some people are opposed to it on the grounds that there are people of color who come from well-off backgrounds who need help much less than some disadvantaged white people. Or, when arguing about school backgrounds and university entrance, some people argue that there are state schools (say, famous grammar schools in the UK) that give people just as much advantage as some private schools, if not more. Should we help those people? I still believe we should try to help all people of color, and all people from less privileged schools, even if some of them don't "need" it.

When arguing about screening for cancer, some people are concerned that the tests are not entirely accurate and result in some positive results even when people don't have cancer. This causes them unnecessary trauma and sometimes unnecessary treatment. This is a serious concern, but I still think this is better than people having cancer that is undiagnosed for too long, making treatment difficult or impossible.

When arguing about sexual harassment, some people are concerned that if we take all accusations seriously we will end up with people (usually men) suffering from the stigma of accusations even if they're innocent. However, up until now we have a huge problem of too many people getting away with sexual harassment, sexual assault and indeed rape, and thus widespread sexual misconduct being pervasive across society. Being falsely accused of sexual misconduct is indeed a trauma that nobody should have to go through, but I believe we need to be more concerned with the quantity of sexual misconduct going unstopped.

There is something analogous between all these situations although they range across many different aspects of life. The analogy may be implicit but I have found it helpful to isolate it and express it explicitly. At first sight it might appear that only the first two scenarios have something in common:

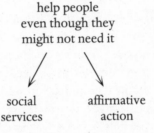

help people
even though they
might not need it

social   affirmative
services   action

The second two might appear analogous separately, because of a different principle:

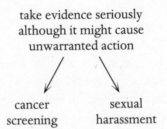

take evidence seriously
although it might cause
unwarranted action

cancer   sexual
screening   harassment

but at a further level of abstraction I can encapsulate all the scenarios in the fact that I seem to believe that false negatives are more important than false positives. Opposite is the diagram of the different levels of abstraction producing different analogies:

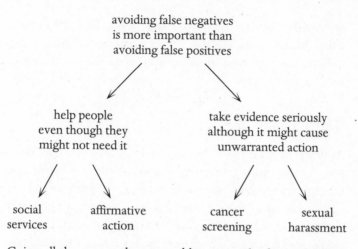

Going all the way to the top enables me to clarify my thinking about complex issues, by distilling the thought process behind my beliefs. I can then apply it to more situations, explain it succinctly to others, and hold it in my brain more easily and thus reason with it better.

Indeed after going up to that level of abstraction I realized I could encompass another situation: the idea of compulsory voting in general elections, as in Australia. I used to disagree with this principle, as I think democracy means that everyone should have the *right* to vote, not the *obligation*. However, I then read an article explaining that it's not about forcing people to vote, it's about forcing the government to make it possible for everyone to vote, to reduce voter suppression and disenfranchisement. I had not thought of that and immediately changed my mind. I now see that it is another example of false positives vs false negatives. Without compulsory voting, you risk false negatives, that is, people who are unable to vote for logistical or more nefarious reasons (such as voter suppression). With compulsory voting you risk false positives, that is, forcing people to vote who don't want to–but they can still leave their ballot blank or spoil it. It is yet another situation where I care about preventing false negatives the most; I just hadn't realized that was the issue until someone pointed it out to me. I will later discuss the fact that I think the ability to change your mind in the light of new information is an important sign of rationality.

TESTING PRINCIPLES

Analogies also enable us to test our principles. We may think that we are doing something because of some fundamental principle of ours, but if that is really true we should be able to move to an analogous situation and apply the same principle. If that doesn't hold it is a sign that our principle wasn't a true one, or that we picked the wrong level of abstraction. Unfortunately people often do this wilfully to try and convince themselves or others they are working to strong fundamental principles rather than prejudice.

For example, perhaps a woman is not picked for a job and the hiring panel is accused of sexism. They may claim that it was not sexism, it was simply that the woman did not have enough experience. However, if they then hire a man who has even less experience, it shows that this principle was not a true logical principle at work. It is often more tricky than this as the analogous situation might not be real, and we are then pushed to imagine what *would* happen in an analogous situation. This question arose in relation to Hillary Clinton and Barack Obama in their respective elections. People who didn't support Clinton were accused of sexism. They fought back and said they didn't support her because "she was a liar" (for example). And yet, many (perhaps all) male politicians lie about things, and are still supported. Similarly people who didn't support Obama were accused of racism. Those who were accused fought back and said they didn't support him because "he was inexperienced", for example. And yet white men with far less experience might win their support.

We can clarify this using diagrams. Someone might think

they're applying a principle about inexperienced people in general, regardless of the fact that person *A* is a woman:

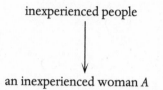

inexperienced people

an inexperienced woman *A*

However, if that is really the case we should be able to pivot, using the abstract principle, to an analogous situation with an inexperienced man:

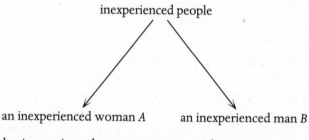

inexperienced people

an inexperienced woman *A*   an inexperienced man *B*

If the inexperienced man gets more credit or more support, then the two are not being treated analogously according to this particular principle, and we should consider whether there is another principle lurking:

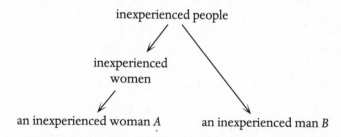

inexperienced people

inexperienced women

an inexperienced woman *A*   an inexperienced man *B*

According to the intermediate level of "inexperienced women", the bottom two items are no longer analogous. The abstract version is this:

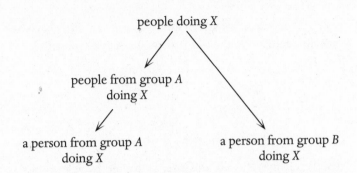

If the person from group A is treated differently from the person in group B, it's a sign that the intermediate principle is at work, not the general one. As discussed in Chapter 3, this applies every time a black person is shot by police in the US, when the same argument repeats about whether it was racism or not. We can ask ourselves whether a white person in the same situation would be treated the same way. If not, then it's a sign that the intermediate principle (the person was black) is at play, not the general one of just the fact that they were doing X.

We should use these principles to test our own arguments as well as those of other people. In arguments about science vs religion there is an analogy that I think scientists should feel uncomfortable about. Many scientists are disparaging about religion because it consists of people believing things without scientific evidence, because of something written in a book or told to them by a religious leader. However, science is asking people to believe in it in a somewhat similar way: there may well be evidence to back up scientific findings, but non-scientists are not expected to read all the research data and check all the research for themselves. Scientists seem to be asking non-scientists to believe them, or believe what is written in books and journals, in an analogous way to religious leaders asking people to believe what they preach, or what is written in the Bible or another holy book. There may well be something crucially different about those situations, but if so we should admit that at this level of abstraction the two appear somewhat similar, and thus, the argument that you "shouldn't just believe what someone says" is not a very convincing argument in favor of science. Using our diagrams again, we have a proposed argument:

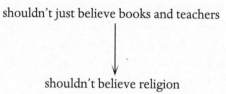

But if this is really a general principle, it is applicable to science as well:

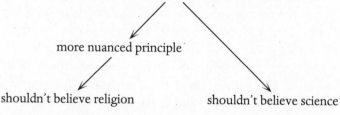

If scientists want to argue that science is more reliable than religion, they need to find a more nuanced intermediate principle that distinguishes the two, and then only go up to the level of that nuanced principle so that science and religion are *not* analogous:

shouldn't just believe books and teachers

more nuanced principle

shouldn't believe religion   shouldn't believe science

That nuanced principle might be that we shouldn't just believe books and teachers unless they are backed up by reproducible evidence, but that still leaves the question of how we can tell if the evidence is reproducible. One might say that religious scripture is eyewitness testimony from people we can't ask, but we might also say that some science is based on eyewitness testimony from people we can't ask, scientists who ventured into the jungle and observed creatures now extinct, or who landed on the moon and reported what they saw. The situation is more subtle than most arguments about it betray, and I think this is a more productive way to see why so many people believe in religion rather than (as some scientists do) declaring that all those people are stupid.

ENGAGING EMOTIONS

Analogies can help us to engage our emotions, if we can find an analogous situation that resonates more closely with us. This is an important way that we can find an emotional connection to back up a logical argument, if the logic by itself has not convinced us or someone else. This can help us understand other people's points of view, or help us explain our points of view to other people.

For example, sometimes men get exasperated about sweeping statements made about men, calling them privileged, aggressive, insensitive, or if we say that there is a "pervasive culture of sexual harassment by men against women". My initial instinct is to argue that we are not actually saying that every man is like that, and also, that when it is the oppressed group (women) calling out the dominant group (men) it is more excusable.

However, I see this with more empathy if I think of an analogous situation in which I am privileged. For example, some people generally regard graduates of Oxford and Cambridge as snobs who were just born to posh families and who are handed success on a plate without having to work hard. I take issue with that, personally, as I think I have worked very hard to achieve what success I have achieved, but I have to be circumspect and realize that I am privileged to have studied at Cambridge and that those who did not have such an opportunity may have reason to feel hard done by in general, by comparison.

The analogy between the situations both helps me understand why men are frustrated, and also helps me understand why people feel animosity towards Oxbridge graduates. This analogy uses the abstract concept of a power relationship, which I will depict with this symbol ▽ as follows:

powerful group

▽

oppressed group

I can then use this to pivot between a situation in which I am in the top group and a situation in which I am in the bottom group (shown in bold in each case), and thus understand both better.

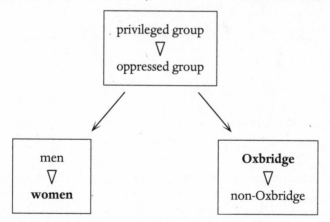

Similarly as an Asian person I can pivot between being in the oppressed group (see over):

white people

▽

non-white people

and being in a privileged group in the context of non-white people (as Asian people are arguably more privileged than others among non-white people)

Asian people

▽

black people

I can use this analogy to perform a pivot:

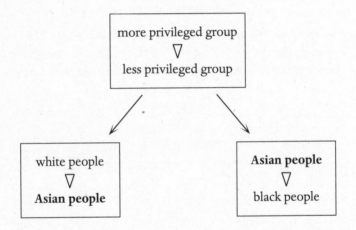

and understand racial discrimination from opposite points of view. To abstract from this further, it comes down to the fact that everyone is less privileged than someone, and more privileged than someone else:

group *A*: more privileged than you

▽

you

▽

group *Z*: less privileged than you

So everyone could perform a pivot to see things from both ways:

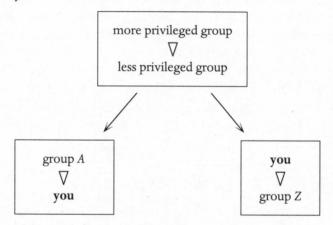

As discussed in Chapter 6, sometimes people are liable to see *themselves* only in the less privileged situation, and see *other people* in the more privileged. It is offputtingly hypocritical when someone complains about their treatment at the hands of their group *A* while simultaneously inflicting analogous treatment on group *Z*. This happens when white women complain about sexism but exclude or neglect women of color, or when white gay men complain about homophobia but exclude gay men of color.

It is understandable to be more aware of people more privileged than you as they are the ones most likely to be a threat to you or prevent your progress, but we need to learn to perform more of the above pivots and become more aware of those less privileged than us without feeling that it contradicts or invalidates whatever lack of privilege we also experience.

WAKING US UP WITH AN EXTREME

Analogies can also engage our emotions not by finding a situation closer to our own lives, but by pushing a principle to an extreme to shock us into seeing that a principle is not so fundamental after all. For example, some people say that universal healthcare is bad because everyone should take responsibility for themselves and not expect someone else to, say, help fund their healthcare.

But in that case does that mean everyone should also take responsibility for protecting themselves and thus we should have no police? No military? No public transport? Should we have no team sports? No families? No basic infrastructure like roads?

Here is the diagram of the supposed principle:

everyone should take responsibility
for themselves

↓

oppose universal healthcare

Here is an extreme that is "analogous" according to the supposed principle:

everyone should take responsibility
for themselves

oppose universal healthcare   oppose roads

If they do in fact believe in roads, then a more nuanced argument is needed to explain why they oppose universal healthcare.

Sometimes someone will counterargue that pushing it to that extreme makes it different. This may well be true, but then the general principle is not really a general principle – it only works within some limits and then, typically, we can discover that disagreements are about where those limits are, rather than about the principle itself. And there is probably a gray area in which the principle gradually stops working. The question about healthcare should really be one about *the extent to which* we believe people should look after themselves and the extent to which society and government should look after people. Political disagreements often come down to a fundamental difference in basic axioms, concerning just how much responsibility someone thinks a government should take, compared with individuals. Other times it is about what counts as a necessity and what counts as an optional extra. We might then get this diagram:

everyone should take responsibility
for themselves

everyone should take responsibility
for their own optional extras

oppose universal healthcare   oppose roads

This actually explains the sense in which the healthcare denier thinks healthcare and roads are different. Then we can argue about whether healthcare is an optional extra or not – another gray area. There are arguments within that about which aspects of healthcare count as essential and which count as extra, with contentious issues including cosmetic surgery, gender reassignment surgery, expensive cancer treatment, IVF, and even basic maternity care.

The purpose of pushing something to extremes is to show that many (or even most or all) general principles have limits to their scope, and the difficult part is not in establishing the principle but in establishing the scope. It is a key to understanding disagreements, as the source of the disagreement is often exactly where to draw the line, rather than the principle itself. It's a way we can use gray areas, to show that the difference between opposing positions might not be black and white, but in shades of gray. If we can show that a difference in positions is quantitative rather than qualitative, we have started bridging the gap between opposing ideas.

PICKING THE RIGHT LEVEL OF ANALOGY

My wise friend Gregory Peebles says analogies are like bridges that can take us *anywhere* – so we'd better be careful what bridge we choose. Indeed, if we pick a high level of abstraction we'll encompass practically everything as analogous, which could include things we didn't want at all. This is how using analogies can sometimes go wrong and cause worse arguments rather than better ones.

Using an analogy in a discussion usually goes like this. You're trying to argue or explain a statement A. You draw an analogy to a statement B that is more engaging, accessible, or clear cut. Implicitly there is a principle X at work. The claim is:

A is analogous to B.

B is true.

Therefore A is true.

This is much less watertight than using an actual logical equivalence:

A is logically equivalent to B.

B is true.

Therefore A is true.

The reason is that the analogy must go via some implicit principle *X*. The unspoken part of what is going on is:

> *A* is true because of principle *X*.
> *B* is also true because of principle *X*.
> *B* is true.
> Therefore *A* is true.

There is now a logical flaw in the argument, which is that just because *B* is true it doesn't mean that principle *X* is true. In a sense we are trying to move *backwards* up the right-hand arrow.

This is made worse by the fact that in normal arguments in life we often don't state what principle *X* is at all: we just leave people to infer it from statement *B*. This is very flawed because there are many possible principles that could play the role of *X* and they could give us very different results. We gave an example earlier with different types of minorities and whether or not their experiences are analogous. This can become a more concrete argument if it is to do with what should count as acceptable or morally allowed behavior.

For example, supporters of same-sex marriage say that a same-sex relationship is no different from a heterosexual one so same-sex couples should be allowed to get married. They are using this general principle:

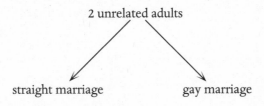

Those opposed sometimes claim that if we allow same-sex marriage, then "the next thing we know we'll be allowing incest". They are erroneously (or wilfully) supposing this level of abstraction is at play (see over):

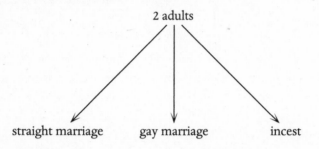

2 adults

straight marriage   gay marriage   incest

The disagreement in this argument is in the principle X that is being assumed to be the cause of A and B. Those opposed to same-sex marriage want the principle only to go as far as "unrelated man and woman" so that straight marriage is not analogous to gay marriage:

an unrelated
man and woman

straight marriage   gay marriage

In seeing someone adopt a higher level of abstraction their imagination or fear sends them escalating much further than the other person has really gone:

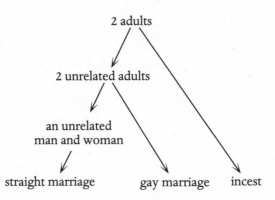

In fact, there's a big hierarchy of increasingly wild arguments. Some people think allowing gay marriage will encompass paedophilia or bestiality. If we make explicit the abstract principles behind these different concepts, we get a diagram like the one below. Each arrow represents the process of going from a principle to an instance of the principle:

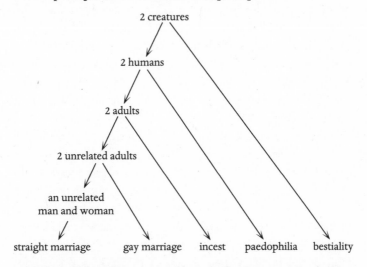

We see that as we get more and more abstract, more and more extreme examples get included. The fact that someone is pro–gay marriage does not mean they have necessarily accepted the principles all the way up to the top.

It is worth remembering that there used to be further levels *below* "an unrelated man and woman", when marriage was not allowed between white people and non-white people. A more nuanced argument than "the same" vs "not the same" would be to think about where on that left-hand edge is a justifiable place to stop. Claiming that going up one level necessarily involves going up more than one is an erroneous argument.

IMPLICIT LEVELS

Many of these problems arise because in life, unlike in math, we do not explicitly state what abstract principles we are referring to, leaving it to be inferred from the analogy. But the people who hear the analogy can infer the abstract principle in different ways, and are especially likely to do so if they disagree with us.

In a way the most reasonable abstract principle to infer from an analogy is the *minimal* one, like the lowest common multiple, or the first meeting place of the arrows in the diagram. In the above example, the higher meeting places were much too high – they weren't minimal and it was not reasonable to assume that the pro-gay marriage person believed in the generality any further than the lowest meeting point of "2 unrelated adults".

One of the reasons analogies are ambiguous in normal life is that we rarely make explicit what abstract principle we are invoking. After all, one of the main points of the exercise is to appeal to people intuitively and *not* have to exert their abstraction abilities. However, everyone then has to guess what principle is being invoked. Whereas in math, making the abstract principle explicit is practically the entire point. We look at analogous situations A and B, and make a precise statement of what principle X we are going to study that causes them. There can therefore be no ambiguity: if A and B are both examples of X, and X is true, then A and B must both be true.

Disagreements over analogies basically take two forms, as in the argument about gay marriage. It starts by someone invoking an analogy of this form:

However, typically *X* is not explicitly stated. Now someone objects, either because they see a more *specific* principle *W* at work that is really the reason behind *A*, so they do not consider *A* and *B* to be analogous:

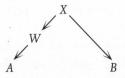

Or, they see a more *general* principle *Y* that they think the first person is invoking. This makes some other thing *C* analogous, and they object to that:

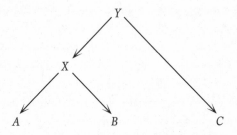

In both cases it would be more helpful to be clearer about what the principles at work really are, to explore the sense in which the different cases are and aren't analogous, rather than just declaring that something is or is not analogous.

For example, is racism by white people against black people the same as racism by black people against white people? Here is a diagram that shows the sense in which it is and isn't the same:

```
              prejudice of people
                against people
                  ╱        ╲
                 ↙          ╲
     prejudice of privileged people    ╲
        against oppressed people        ╲
               ╱                          ╲
              ↙                            ↘
   racism of white people        racism of black people
     against black people          against white people
```

The argument is really about which level of principle we should go up to.

There is no right answer to what a good level of abstraction is for understanding a situation. All analogies break down somewhere. That is the whole point of an analogy: it is not the same as the original thing; it is similar in some way but therefore it is also different in some way. Pointing out that an analogy breaks down does not mean the analogy is bad. But if the analogy breaks down in some way that is relevant to what we're discussing, that could be more important.

I think the best we can do is explore different levels and find out what levels cause analogies to appear and to break. This shows us what abstract principle is at work. In the end the whole aim is to reach greater understanding of in what way situations are equivalent and in what way they are not. This is the subject of the next chapter.

14

EQUIVALENCE

WHEN THINGS ARE AND AREN'T THE SAME

ONE LONG-STANDING MYTH about mathematics is that it is all about "getting the right answer". That everything is simply right or wrong. Another pervasive myth is that it's all about equations.

These myths both have some truth in them, but are very far from being completely true. Equations do arise a lot in school math, but as the objects we study become more interesting than numbers, the issues become more interesting than equations too.

But there is an important idea at the heart of equations, which is that they are about finding things that are the same as each other. However, nothing is actually the same as anything except itself. As we saw in Chapter 8, equations are all lies apart from those of the form $x = x$, which is not at all illuminating. Other equations have some truth in them, so, *there is a sense in which* the two sides are the same, but, crucially there is a sense in which they are not. We saw that in this equation: $10 + 1 = 1 + 10$ the two sides *are* the same in that they produce the same answer, but they are *not* the same in the sense that technically they describe a different process.

The point of equations in math, where they do arise, is to find two things that aren't the same in one sense and that are the same in another sense. We can then use the sense in which they are the same to pivot between the ways in which they aren't the same, and thus gain some understanding, as we described in the previous chapter. This idea continues into research-level math, where the senses of "sameness" become more and more subtle, and increasing amounts of technical effort have to be put into finding and describing appropriate notions of sameness.

We saw in the previous chapter that analogies involve finding situations that aren't exactly the same, but that are the same in some sense – and necessarily different in another. We can then pivot using the sense in which they're the same, and perhaps land in a place that is logically similar but more emotionally engaging, or more extreme and thus easier to judge morally.

But there are different levels of abstraction that produce different analogies with different quantities of sameness. Which one should we pick? This is another place where gray areas arise all the time. There are different ways in which things can be equivalent and not equivalent. A better question than "Are these the same or not?" is "In what sense are these the same and in what sense are they different?"

EQUIVALENCE IN MATHEMATICS

When we first learn math it is all about numbers. And numbers don't really have ways of being the same as each other except by being equal, so math is mostly about equations as well. However, as math progresses it starts involving things that are much more interesting and subtle than numbers, like shapes and curves and surfaces and spaces and patterns. These have many more ways in which they can be the same as each other, depending on how stringent you want to be. As humans we are quite used to different levels of sameness in different situations. For example, if two different people write the letter "a", it won't come out exactly the same, but we'll recognize both as being "the same letter". However, we might be able to tell they're written by different people. If I write the letter "a" several times they'll all look slightly different, but a handwriting expert should be able to tell that they're all written by the same person:

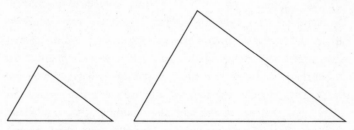

One of the problems with "handwriting" fonts on computers is that if you look closely you can see that every letter "a" is exactly, precisely the same, so while the text might look handwritten at first glance, on close inspection you can quickly tell it isn't.

In math too we may want different levels of equivalence for different contexts. You might remember that two triangles are called *congruent* if they are exactly the same shape and size, that is, they have the same angles as each other *and* the same length of sides. They are, in some sense, *exactly* the same. If they have the same angles as each other but possibly different lengths of sides then one is a scaled version of the other and they are called *similar*:

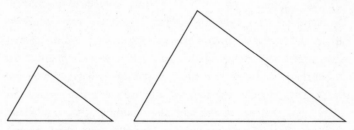

They are not exactly the same, but there is a sense in which they are the same. In a way the second one is what the first one would look like if we held it closer to our eyes. And what about this one?

It's just the first triangle but flipped over sideways. Does flipping a triangle over make it a different shape? It depends what you're using it for. When children learn how to write letters they sometimes have trouble getting them the right way round. I sympathize, because we are expecting them to understand that these are not the same:

$$S \quad Ƨ$$

although in some very strong sense they are the same shape.

In fact, all equivalence and sameness depends on what you take into account and what you ignore – except strict equalities like $x = x$. There is no way to make that not equal. So as math becomes more advanced it becomes more and more about finding *the sense in which* things are the same and *the sense in which* they are different. The more dimensions something has, the more different ways there are in which it could count as the same, and the more subtle it becomes.

One of the most famous recently solved mathematical problems was the Poincaré conjecture, which was essentially about this subtlety. It involves looking at higher-dimensional spaces (of a certain special kind) and asking what happens if we count them as the same if they're "the same shape" where we don't care about size but we also don't care about curvature and pointiness. This can be thought of as taking a playdough shape and squashing it around without breaking it or sticking parts together, and counting that as the same, with the famous example that in this case a (ring) doughnut is "the same" as a coffee cup with a handle – the hole in the doughnut corresponds to the handle in the coffee cup. Of course, it's a bit hard to understand what playdough would be in higher dimensions, but that's why math doesn't actually refer to it as playdough. Anyway, the Poincaré conjecture is about which spaces count as the same and which ones count as different under this condition. It doesn't mean that those spaces *are* the same; it just means that viewed in this particular light they can be seen as the same.

Under this type of sameness (which is technically called "homotopy equivalence") a square is the same as a circle. But a square is certainly not the same as a circle if, say, you are making a wheel. A square might be as good as a circle if you are making a cake – except that a square cake tin is harder to wash than a circular one because of the corners. (And also the corners are likely to get more cooked than the rest of the cake.)

In life there are even more ways in which two things could be considered the same and not the same, because the things we think about in life are much more subtle than the things we think about in math. So really, we should be even more careful about the *ways in which* things are the same and different, rather than just declaring they are the same, or not the same. In fact, it's another example where we tend to see in black and white instead of comprehending the whole gray area in between.

We have seen the most obvious sense in which things are the same: if they are actually equal. But this is not helpful. At the other extreme we can see the most extreme sense in which things are not the same – false equivalence. We will discuss that next, before thinking about the gray area.

FALSE EQUIVALENCE

Periodically an argument breaks out about the "gendering" of children's clothes and toys. On one side, people point out that boys and girls can play with the same toys if they want, and there's no reason that a T-shirt with dinosaurs on it should be called a boys' T-shirt, and one with flowers a girls' T-shirt. Usually people on the other side will complain about political correctness and declare that we should just "let boys be boys and girls be girls".

It seems to me that they are equating the argument "boys

and girls can play with the same toys" with a desire to turn boys into girls and girls into boys. This is a false equivalence.

Erroneously equating one statement with another one is a devious tactic, and is not logical. It is often invoked to twist someone's words and turn their reasonable position into a much worse one, and then attack them for it. It often drives arguments into further and further extremes. It involves an incorrect piece of logic, claiming two things are logically equivalent when they are not, so is a form of logical fallacy.

In Chapter 12 we talked about good logical ways of dealing with gray areas. One way that false equivalence arises is when we don't deal with gray areas well – we often end up lumping things in with extremes. "If you're not with us you're against us." Well, you might be neither totally with them nor totally against them. You might support some things someone does but not others. Some of these are versions of false equivalence, and some can find their root in false negations, when a statement is negated incorrectly to produce a more polarized opposite. Sometimes it's a flawed analogy as described in the previous chapter, where someone goes to too high an abstract level and claims your statement is analogous to something absurd. In all cases this creates divisions instead of finding common ground.

In fact, I have somewhat pushed the argument about boys' and girls' clothes to extremes myself: in order to show the counterargument in the worst possible light, I have picked the most harmless manifestation of the "politically correct" argument that I can think of. Really there is a more complex argument going on underneath, about gender stereotypes and gender pressures. When people get very upset about someone else's point of view it is worth trying to work out what the underlying issue really is, which might not be logical at all. It might be very personal.

PERSONAL TASTE

When someone expresses personal taste, it sometimes happens that other people get offended. I might say "I hate toast" (which I do), and someone else will take offence because they really like toast.

This might sound silly with the toast example, but perhaps it's more believable that when I say "I find Mozart boring" people who like Mozart take it as an insult, or if I say "I don't like jazz" people who like jazz think I'm criticizing them. Or if I say "I don't want to be fat" and people think I'm fat-shaming.

I think a false equivalence is at work. Someone hears

"I don't like toast."

and thinks it is equivalent to

"I don't like people who like toast."

This is a false equivalence: the two statements are not logically equivalent. I am perfectly happy for you to like toast, I just happen not to like it myself. This is different from

"I don't like stealing."

because in fact I also don't like people who like stealing. There is also a false equivalence between

I don't want to be fat.

and

I think people who are fat are bad.

This is not a logical equivalence. We could make an abstract version of this:

I don't want to be X.

and

I think people who are X are bad.

If X is "a doctor" then it's obviously a false equivalence: I don't want to be a doctor, but I don't think doctors are bad, I just happen not to want to be one. On the other hand if X is "a horrible person" then the second statement actually explains the first one: I think horrible people are bad, so I don't want to be one. This shows that the error is a form of converse error. We have these statements:

A: I don't want to be X.

B: I think people who are X are bad.

This implication is not very contentious:

$$B \implies A$$

but the people getting upset with me are mistakenly thinking this converse is true:

$$A \implies B$$

If the converse *were* true it would make A and B logically equivalent, but the false converse causes the false equivalence.

A similar form of false equivalence is when I say "I like weighing myself every day" and someone thinks it is equivalent to "I think everyone should weigh themselves every day", thus they get offended if they don't do that. Again we can look at

I like doing X every day.

and

I think everyone should do X every day.

For example, what if X is "playing the piano"? I certainly like playing the piano every day but that doesn't mean I think everyone should do it. On the other hand if X is "brushing my teeth", I like doing that every day and think everyone should do it every day if they can.

So much for the logic of the situation. It is an example of

how I can always be right, by restricting my statement to my own personal taste or aspirations about myself.

However, people do often get upset when I say I don't want to be fat, and they are rarely placated by my explanation of the logic. In the next chapter we are going to talk more about attempting to override emotional responses, but I think it's worth noting here that uncovering the false equivalence gives us the logic behind their feeling of offence: they are equating me not wanting to be fat with me criticizing people who are fat. A more compassionate thing to do with this piece of logic is to admit that human language is not the same as logic, because it can carry connotations in a way that logic doesn't. If I am genuinely not criticizing fat people, I should perhaps find a different way to express myself. Or I should examine myself carefully to see if a tiny part of myself is criticizing them but hiding behind the logical security of making it a statement about myself, like when people express an inflammatory view and claim they are playing devil's advocate.

DAMNING ACCUSATIONS

We have just seen how false equivalence causes people to think I am accusing them of something when I'm not. Another case is when false equivalence causes people to make damning accusations of me. This quickly spirals into an antagonistic argument, the kind where instead of trying to see eye to eye, two people try to cast each other in worse and worse lights and thus drive the disagreement further and further apart.

For example, I might say that I don't want to be fat so I am careful what I eat. Someone might then say to me "You're fat-shaming and that is misogynistic". I get criticized for not wanting to be fat in many ways, but this is one of the common ones. Apparently not wanting to be fat makes me misogynistic. Suddenly it turns into an argument about misogyny.

A sign that someone is about to make a false equivalence is

when they launch in with "You're basically saying. . . " which is a sign they're about to twist your words into something they're not. You might say "I think it's unfair that some people just inherit tons of money so never have to worry about anything, whereas other people are born with nothing." Someone might argue "So you're basically saying that everyone's money should be confiscated when they die so that they can't pass it on." That's not what I'm saying at all. There are a lot of possibilities in between simply allowing inequality to be handed down from generation to generation and forcibly confiscating all money at death. But the arguments in between are complicated. They involve things like trying to even out inequality at all stages of life, trying to help those who are born with nothing so that they can rise out of that situation and then maybe they too can have something to pass on to their children. The arguments in between are usually too complex for a fast-paced shouting match or an online exchange.

In general, an antagonistic argument driven by a false equivalence might go like this:

You are saying A.
A is equivalent to B.
B is bad.
Therefore you are a terrible person.

The logic of this can fail in two places: maybe B isn't really bad, or maybe A isn't really equivalent to B. Of course, it could also fail in both places. Arguing against logic that fails in multiple places can be strangely confusing, like you're giving too many excuses or protesting too much. An example of this is if you support reproductive education at school, and someone declares that's equivalent to condoning sex outside marriage, so is evil. I don't think it is equivalent to condoning sex outside marriage, but I also don't think sex outside marriage is evil.

FALSE DICHOTOMY

There is another aspect to the argument about boys' and girls' toys: in fact it is a false dichotomy. The arguer thinks there are only two options:

A: Label some toys "for boys" and others "for girls".

B: Force boys to be girls and girls to be boys.

A dichotomy is when the options are cleanly split between some option A and option B, with those being the only two possibilities. A false dichotomy is when you *think* the options are perfectly cleanly split between A and B, but in reality they are not. As a result you mistakenly think that A is equivalent to "not B" and B is equivalent to "not A", which is why it's an example of a false equivalence.

In the above argument, the person thinks that B is the negation of A, but it isn't. Not doing A could involve simply removing the labels on toys that declare them to be for one gender or the other; boys can still be boys and girls can still be girls and everyone can play with whatever toys they want. I suspect the people who believe in the false negation are acting out of some deep fear of gray areas, but it's hard to pin down exactly what it is. A simpler case is if you say to someone "I wouldn't call you skinny" and they burst into tears wailing "You think I'm fat!" They are jumping to this false dichotomy:

A: I am skinny.

B: I am fat.

possibly out of fears that come from society's pressure on women to be thin. It's a false dichotomy because it's possible for neither A nor B to be true.

We can express this in the pictures shown on the next page, showing the relationships between the different statements involved. Here is a true dichotomy, where the circle is perfectly split between A and B:

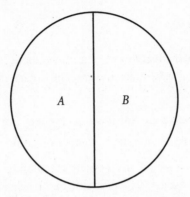

False dichotomies can then occur in two ways. Sometimes it's a false dichotomy because it's possible for neither thing to be true, as in the case of being skinny vs being fat:

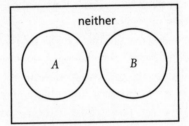

Sometimes it's a false dichotomy because it's possible for *both* things to be true at the same time:

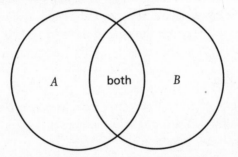

Of course, it's possible to fail in both of those ways at the same time. We'll see an example of the second type of false dichotomy now.

DIETING

A false dichotomy that often gets me into trouble with people is:

A: Some people should watch what they eat (because it helps them stay healthy).

B: Some people should *not* watch what they eat (because it hinders them).

It is certainly helpful for me to watch what I eat, but I acknowledge that for other people it causes problems. We should all pick whatever's best for us. Unfortunately when I say A, people think I'm negating B, and then they get angry and say that it is better for them not to "go on a diet". But this is not a disagreement: it is perfectly possible for both A and B to be true (and I'm sure they are both true).

The true negation of B would be the much more extreme statement "Everyone should watch what they eat", so perhaps the error is a false equivalence between this and my rather mild original statement.

We saw in Chapter 6 how helpful it can be to draw diagrams of relationships between concepts, so I will now draw some for false dichotomies. In the following diagram, my point is the top left one, and the typical refutation is the bottom right argument, when in fact these do not disagree:

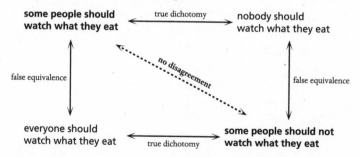

The really funny thing about this argument is that it usually turns into a meta-argument about whether or not we're disagreeing. I try to point out that we're both making the same point, and the arguer usually insists that we're not. I think they really feel that I am criticizing them for not being careful about what they eat. The result is that we go from these two compatible reasonable points:

A: Some people should watch what they eat.

B: Some people should not watch what they eat.

To these two absurd and antagonistic ones:

A: Everyone should watch what they eat.

B: Nobody should watch what they eat.

This is the other diagonal of the diagram:

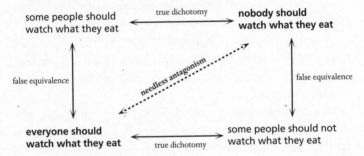

I think this is another example of the general false equivalence between making a choice A, and thinking that no other choice is valid. Just because I choose A, it doesn't mean I think everyone should choose A. And yet when I choose something that someone else didn't choose (like watching what I eat) people too often assume I am criticizing their choice. I suppose in many cases people *are* criticizing different choices, but it doesn't have to be that way, if we don't let false dichotomies push us to extremes.

STRAW MAN ARGUMENT

False equivalences are a source of straw man arguments, where an argument is replaced by one that is much easier to knock down (a straw man) and is then duly knocked down. However, if the new argument is not equivalent to the original one, all you have done is knocked down an argument that nobody was making.

The emphasis on "STEM" subjects (Science, Technology, Engineering and Mathematics) is sometimes criticized on the grounds that creativity is important too. This is making a false dichotomy between science and creativity, possibly stemming from a more fundamental one between

A: Being creative.

B: Being logical.

Sometimes this causes creative people to justify being illogical, or to refuse to be logical. There's also a false equivalence between creativity and art, and another between logic and science. There is logic and creativity in both art and science:

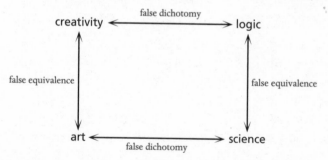

Thus to use creativity as an argument against emphasizing STEM education is a straw man. There are plenty of valid ways to advocate for arts education without denigrating science.

The most pernicious straw man argument I see regularly is "All lives matter" as a response to "Black lives matter". The straw man as I see it is this. What the slogan "Black lives matter" really means is "Black lives matter just as much as other lives, but are currently being treated as if they do not matter as much, and we need to do something to correct this injustice." Of course, that is not as catchy a slogan.

The straw man argument wilfully misinterprets "Black lives matter" to mean "Black lives matter and other lives don't" which is easily refuted by saying "All lives matter". The frustrating thing about this argument, apart from the fallacy of it refuting something that nobody was trying to argue, is that it is almost impossible to argue against it. To do so we would have to argue "Some lives don't matter", and apart from some reprehensible extremists, we do not think that. As in the rather more trivial example of watching what we eat, we can represent this push to extremes in diagrams. Strictly logically, there is no disagreement between the statements "black lives matter" and "all lives matter". However, if we replace "black lives matter" with the inequivalent straw man argument "some lives don't matter" then the diagonal with no disagreement in the first diagram (below) gets pushed to the extreme and antagonistic diagonal of the second diagram (opposite):

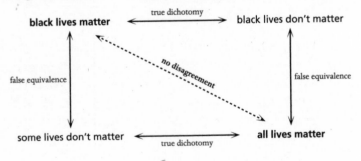

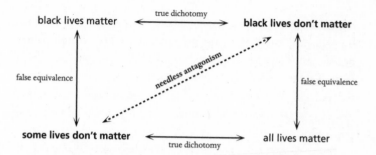

In order to have a logical argument, we need to persuade the person who opposes "black lives matter" to refute the true argument "Black lives matter just as much as other lives, but are currently being treated as if they do not matter as much, and we need to do something to correct this injustice."

This has three parts, connected by "and":

A: Black lives matter just as much as other lives.

B: Black lives are currently being treated as if they do not matter as much.

C: We need to do something to correct this injustice.

The argument being made is:

$$A \text{ and } B \text{ and } C.$$

To refute this, you just need to refute one of those. If the refuter is persuaded to admit which one (or more) they disagree with, we can get closer to understanding what the argument is.

If the refuter does not agree with A then we can conclude that they are an out-and-out explicit racist. If the refuter does not agree with B then I would conclude that they are ignorant about the state of the world, or deluding themselves. If they do not agree with C then they might not think they are actively racist because they might not be engaging in active oppression, but if you do not try to take steps to defeat oppression then you are arguably complicit with the oppressors. At least if we gain this clarity we can have a more productive investigation into why the person disagrees with B and C. Is it because they think that black people are bringing it on themselves? Is it because they do not think it is anyone's responsibility to help other people, in general? In the first case I would try to persuade them to understand how systems work, that people do not operate independently of a system that has put them there, and in this case it is a whole historical system of abuse, mistreatment and oppression.

In the second case we are more likely to reach an impasse. If someone doesn't believe in helping other people, that is a basic difference in axioms from mine.

There is also a non-logical but emotionally valid possibility for what the opposer is really trying to say, which is they oppose the "Black Lives Matter" movement on the grounds that they associate it with anger and aggression which they think is counterproductive. In Chapter 15 we will discuss the importance of uncovering the emotional reasons for arguments that don't seem to make logical sense. In this particular case we should then have a discussion about

1. whether or not the Black Lives Matter movement is really synonymous with anger and aggression, and

2. when anger and aggression are reasonable.

We could perform an abstraction pivot to build a gray area bridge and agree that there are some situations in which aggression is not reasonable (for example if you go to buy something in a shop but they've sold out) and some situations in which aggression is reasonable (for example, if someone is trying to kill you then you could be forgiven for fighting them aggressively). The question is then one of gray areas and the possibility that the treatment of black people in the US is similar enough to someone trying to kill them that they can be forgiven for fighting back aggressively.

Of course, this analysis is far too complex for interactions on social media, but unfortunately this is where so many of these arguments take place. It is also too complex for people who are riled up by anger or fear. Complex arguments require a certain level of calm. This is true in mathematics as well – if I get overexcited about a result I'm about to prove, I won't be able to prove it. Or if I panic that my time is running out, perhaps because of a looming deadline or a talk I'm supposed to give, I also won't be able to prove it.

If you want to avoid someone making a straw man argument at you, it helps to state your position precisely. We have seen from Chapter 5 that we should look at all factors that cause a situation, and really if we express ourselves imprecisely and leave ourselves open to an easy straw man argument we are a little bit complicit in it, even if the straw man has involved some wilful misinterpretation by someone determined to disagree with us. It takes longer to make an argument precise than to use a catchy slogan, which is problematic in a world of memes and 280-character messages. In Chapter 16 we will come back to our need for slower arguments despite our fast-paced world.

ANALOGIES

In the last chapter we saw that complex false equivalence situations can arise when you try and argue a point by making an analogy. This is usually a bit fraught because analogies don't always play a strictly logical role in an argument, but rather, an emotional role in persuading people of things. These false equivalences are not exactly logical fallacies any more, but something more subtle in the gray area.

A false equivalence logical fallacy often comes into an attempted logical argument like this:

A is (falsely) equivalent to *B*.

B is true.

Therefore A is true

Someone might well successfully argue statement B, but if B is not actually equivalent to A then they have not argued A at all. Often in life this takes the form:

A is (falsely) logically equivalent to B.

B is good/terrible.

Therefore A is good/terrible.

This general situation also happens in a straw man argument, as we discussed earlier, for example:

Saying we should remove
gender labels from children's clothes (A)

is logically equivalent to

saying we don't want to just
let girls be girls and let boys be boys (B).

B is terrible.

Therefore A is terrible.

Using analogies is a more nuanced way of using equivalence as a pivot. Instead of claiming a logical equivalence between A and B, we invoke an analogy involving a certain principle X, which we represented in this type of diagram in the previous chapter:

We can then try to convince someone of the principle by applying it to situation B that makes a more emotional connection. The idea is that A and B are equivalent at the level of the abstract principle X. The example doesn't justify the principle, but is supposed to help us *feel* the principle.

As we discussed, the question is about the level of abstraction and thus the extent to which A and B really are similar because, after all, everything is the same if we go up to a high enough level of abstraction, but that would not illuminate the argument at all. So the question should not be "Are these things

equivalent?" but rather "in what sense are these things equivalent?"

In the diagram below, A and B are analogous according to
principle X, but not according to principle Y. If someone says A
and B are analogous we can't simply say it's a false equivalence
or not: we should discuss which principle is appropriate.

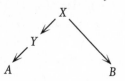

An example of this situation is in arguments about "mansplaining". I have found that using the term "mansplain" is likely to
cause at least one person (usually a male person) to get rather
wound up, but I usually think that is more evidence that we
need to continue using it.

Mansplanation is not just when men explain things to
women in a patronizing way. This is a false equivalence.
Mansplanation is in fact when a man explains something to a
woman despite the fact that there is strong evidence that the
woman already knows it, and a man is ignoring this evidence as
part of the systemic societal assumption that men know more
things than women, whether or not he is consciously applying
that bias in this particular case. This happens to me a lot, for
example, if a man explains something very basic about infinity
to me despite the fact that I have written a book about infinity.
Rebecca Solnit coined the term after a guy explained her own
book to her, after she mentioned having written a book on the
subject. He did it to show that this book was much more
important than hers, apparently without ever considering that it
might *be* hers.

Sometimes it happens that someone explains your own area
of expertise to you, so the evidence that you don't need the
explanation is in the fact that you are an expert. But sometimes
the evidence that you don't need the explanation is that you

already said that very thing yourself. The assumption of the woman's ignorance isn't just generally patronizing, it is specifically part of a broad pattern in society of men undervaluing or neglecting the contributions of women. Because this contextual aspect is built into the word "mansplain", it is by definition not something that women can do. The frequent straw man argument is that "women mansplain things too". I would say the straw man in this case is misrepresenting the argument as claiming that only men explain unnecessary things in a patronizing way. While in my experience it is almost always men who do that, that is not the point. The point is that when men do it, it is part of a society-wide assumption about women, and that is why it is so aggravating.

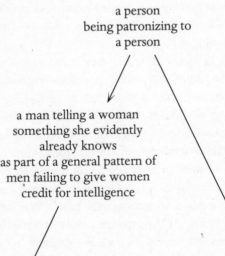

Those who think "women mansplain things too" seem to think that mansplanation was coined just to describe men being patronizing to women. If we go all the way up to the top level of this diagram then this is indeed analogous to women being patronizing to men, but this is a level of abstraction too high.

FALSE FALSE EQUIVALENCE

We've seen with mansplaining that, if we're not careful, false equivalences can be used to shut down valid arguments. But so can false *accusations* of false equivalence.

Imagine an argument about whether higher education should be paid for by the government. One person objects on the grounds that higher education is optional and everyone gets to decide for themselves whether they go or not, so if it's paid for by the state then individuals get to decide how to spend the government's money.

Someone else might argue that this is true of healthcare as well – when healthcare is paid for by the state, individuals get to decide whether to go to the doctor or not, so they too are essentially deciding how to spend the government's money. Of course this argument won't work with someone who doesn't believe healthcare should be paid for either, but what if the person objects to government-funded higher education but supports government-funded healthcare? Is that inconsistent?

At this point if you accuse someone of inconsistency in this way they're likely to exclaim "That's not the same", which is the usual battle cry when someone is trying to argue against another person's analogy. But of course the analogy isn't the same – it's an analogy. The question is whether the analogy breaks down on a crucial rather than an irrelevant point.

You could argue that people don't "decide" to go to the doctor, they just go when they're sick. Whereas in fact I think an arguably analogous level of decision goes into going to see the doctor and going on to higher education. Different people decide to go at different points for different reasons. I don't think it's as clear cut as "hypochondriacs vs normal people", and "sick people vs well people". Some people seem to think it is a black and white logic system:

	well	sick
hypochondriac	doctor	doctor
normal	no doctor	doctor

whereas I think it's more like a fuzzy logic system:

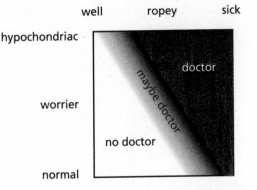

Some people refuse ever to go to the doctor. Will Boast's haunting memoir *Epilogue* talks of his father dying in his car by the side of the road rather than hitting the emergency button, so averse was he to the idea of getting medical help.

Aside from that, there are situations in which most people would seek medical attention, such as if they've broken all their limbs, or got third-degree burns, and others in which some people would go to the doctor and others would decide not to. Notoriously, many people go to the doctor when they have a bad cold, and want antibiotics although antibiotics do nothing for viruses, whereas I just go to bed and drink whisky. All this is to say that we do make choices about using healthcare.

Compare this with higher education. It is not exactly a free choice whether to go or not, as so many professions are inaccessible nowadays if you do not have a degree. This is very different from, say, fifty years ago when it was more of an indulgence to go, unless you wanted to be a doctor or a research scientist (or perhaps a few other things). You could still be a banker, a civil servant, a teacher without a degree. Today you can choose not to go but then you'll probably either be restricted to a low-paid unskilled job, or you'll have to rely on being a rare entrepreneur. Even artists are mostly expected to have degrees now although vocational apprenticeships fortunately still exist in some fields.

The two different points of view can be represented like this. One person thinks education and health care are not analogous, because they are using these principles:

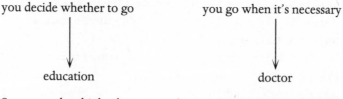

Someone else thinks they are analogous, by using this principle:

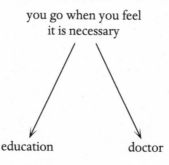

I think with both college and the doctor, most people go because they *perceive* it to be crucial to their future life. However, that level of perception is subjective, and what looks like an involuntary, critical matter to one person can look like an indulgent choice to another. I will concede that there might be some exceptions – rich people who go to university just for the sheer fun of it, as they will never need to depend on their education in their lives. Perhaps this is analogous to hypochondriacs who go to the doctor for no particular reason?

So, are the situations equivalent or not? A claim of "false equivalence" itself needs to be justified. Just saying that something isn't the same does not mean that it isn't *equivalent* in some crucial way.

MANIPULATION

False equivalence and false dichotomies in general lead to false arguments and fabricated division between people who might not really be disagreeing. This can be exploited by politicians, the media, or just people who have more to gain from discord than from unity.

This can be a particular problem in partisan politics, where one party makes opposition their entire purpose. Then if the party on the left moves to the centre to win more votes, the opposition has to move much further to the right in order to oppose them robustly.

Individuals can also cast people, opinions or policies in a much worse light by making a persuasive false equivalence to something considered to be bad. They might declare a position to be "unpatriotic", thereby invoking strong emotions from those people who care about patriotism. For example, some people proclaim that voting to remain in the EU was "unpatriotic", or that kneeling during the national anthem is "unpatriotic".

A nuanced argument would then look into what exactly "patriotic" means. Perhaps it means something like loving and supporting your country. We could then have a discussion about whether believing in the EU means that you don't love and support your country. What if, in fact, you want the best for the UK and you believe that the best means staying in the EU? It then comes back to an argument about what is best for the UK, which is what the argument should be about, rather than simple name-calling.

Similarly for the national anthem, some people argue that kneeling during the national anthem shows that you don't love and support your country. What if, in fact, you want the best for your country and you believe that the best means eliminating racism, and that kneeling during the national anthem is a way of raising awareness and showing solidarity for this cause?

False equivalence arguments can be obstructive, but in worse cases they can actually be destructive. Some homophobic people erroneously perpetuate the myth of an association or even equivalence between homosexuality and paedophilia. Transphobic people make an association between transgender people and perverts. These are abhorrent, demonstrably false equivalences, that can be demonstrated to be false with extensive statistics, but the false equivalences are still perpetuated by people who want to drum up hatred.

These are some of the most destructive ways in which people can be manipulated by emotional means. Such methods of manipulation rely on us not being able to think logically enough to see through them. Unfortunately, because of the power of emotions over logic, there are many people who are either unable to think logically enough, or prevented from doing so once their emotions are stirred up. In the next chapter we will explore what a better interaction between emotions and logic could be.

15

EMOTIONS

WHEN LOGIC NEEDS HELP

EMOTIONS DO NOT LIE. They are never false. If you feel something you are definitely feeling it. If someone tells you that you are not justified in feeling it, that doesn't help. Nor does it help if they tell you that they feel completely differently. There is still always a reason you are feeling it, and in this sense there is always some sort of logic to emotions. Instead of denying or suppressing emotions we should understand and explain them. I will go one step further and say we should even use them: it is important to remember that emotions can, and probably should, play a role even when we're being logical. We use them when doing rigorous mathematics, as we have already seen, and so we should use them when making logical arguments in life as well. Our access to emotions is an important difference between us and computers. Emotions can help us in all of our logical endeavours, and I would even go as far as to say they are crucial.

First of all, emotions can help us work out what we really believe in, just like they help us guess what is logically correct in mathematics before we start trying to prove it. Then when we do start trying to justify things, emotions help us to arrive at logical justifications, if we closely analyse where our gut feelings are coming from.

The next stage of a (useful) logical process involves persuading other people of things. We are going to discuss the importance of using emotions for this. But the emotions shouldn't supersede the logic – they should reinforce it.

Sometimes people try to argue that we should *only* use logic and scientific evidence to reach conclusions. However, if we then meet someone who isn't convinced by logic and evidence, how are we going to persuade them to be convinced by it? We can't use logic and evidence because that doesn't convince them. We are going to have to use emotions.

In a way this means that emotions are much more powerful than logic, and are much more convincing than any other possible method of justification. If you feel something, there is absolutely no way to contradict it. This power should be harnessed in a good way, to back up logic rather than contradict it.

LOGIC AND EMOTIONS

Being emotional does not necessarily equate to being irrational: I think that is a false equivalence. This takes the form of a false dichotomy between

A: Using emotions.

B: Using logic.

I think this is the type of false dichotomy where it is possible to do both at the same time.

Using emotions is not inherently illogical, and using logic is not inherently emotionless either. We can be pushed to a fabricated disagreement in which one person says they believe in using emotions and another says they believe in using logic. But it is possible to do both.

At root this has turned into some needless antagonism about intelligence and sympathy.

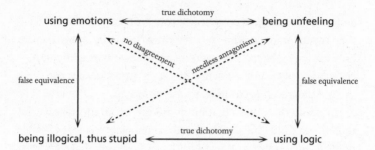

It is possible to be neither emotional nor logical, and it is also possible to be both emotional and logical. That is to say I think it is possible to be in all parts of this Venn diagram:

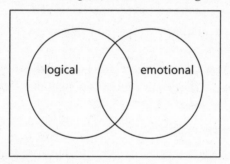

For the left-hand side, there are aspects of mathematics that require one to be very strictly logical. This is not the beginning part, where we are generating new ideas, inventing new language, taking vague leaps in different directions to see what will happen. It is not the end part, where we are writing up our proof for the benefit of other people's understanding. It is in the crucial middle part where we are meticulously proving all the logic steps of our theory to make sure that it is utterly watertight. This part, I think, demands that we be utterly dispassionate, not distracted by what we feel or what we want to be true, so that we can see if logic by itself will hold it up.

For the right-hand side, there are times when it is enjoyable and even beneficial to allow emotions to guide everything. This might be when enjoying sensory experiences, allowing ourselves to be open to art, or just when supporting another person through a difficult or a particularly joyful time. If someone is very hurt it is often no use applying any sort of logic, but the most helpful thing can be to simply sit with them in their emotions and feel things with them.

The only part I can see no good use for is the outside, where one is neither emotional nor logical (although perhaps we should count subconscious actions like reflex actions or automatic behavior like walking or brushing your teeth). By contrast I am going to argue that the most powerful place to be is the place in the middle, where logic and emotions coexist. Not only do they not need to compete with one another, but they can even strengthen each other.

Living too much in the logical world can make it difficult to deal with other people, as they do not usually, or ever, behave entirely logically. On the other hand, those who live too much in the emotional world may have trouble dealing with the world in as much as it *does* behave logically, and does have components and systems that interact with each other. But living very predominantly in the emotional world doesn't mean being actively irrational, it might just mean being guided more by emotions than by logic. And it might mean being unable to follow complex reasoning about the complex world.

Children often live very predominantly in the emotional world. All their emotions are valid and strongly felt, but they are unable to see more complex long-term arguments such as: if you only ever eat ice cream then eventually this will probably not be very good for you. Or even: if you roll around in the snow it might well be fun, but your clothes will get wet and then you'll be miserable.

One aspect of growing up is developing the ability to comprehend longer chains of causation and logic. One concrete way this manifests itself is in the ability to make long-term plans, or make short-term sacrifices for long-term gains, rather than just living for instant gratification in the moment. At least, this is one of my personal axioms; at the other extreme there are some people who strongly believe in only living in the moment, or living entirely emotionally. When adults live strongly in the emotional world it doesn't necessarily mean they are neglecting the logical world. I believe I live very strongly in both. I respect and trust my emotions, but always look for logical explanations of them so that they're not "just" emotions. The two are not mutually exclusive.

EMOTIONS OVERRIDING LOGIC

There are plenty of ways to use emotional responses that do not back up logic, and even to contradict it. One powerful tactic is shock and fear. Once someone is afraid, it doesn't matter whether the thing they're afraid of is real or not. Horror movies are still scary, although they're not real. Clickbait headlines and catchy slogans are also appealing to emotions instead of logic, and too often the logic doesn't hold up. Clickbait headlines often do not accurately reflect what is in the actual article, even if the article itself holds up logically. I read one recently that said "Letter invoking Article 50 declared illegal by judicial review", which was astonishing to me. But when I read the article what it reported was actually that a retired doctor had decided that the letter invoking Article 50 was illegal and was trying to *instigate* a judicial review: very different. It doesn't even matter if you're not British and don't know what the furor over Article 50 is about, it should be clear that "X is declared illegal by judicial review" is not the same as "A retired doctor thinks X is illegal and is trying to instigate a judicial review".

Catchy slogans often sound good but either don't make sense or have no content. "Weight is just a number", people like to say, or "age is just a number". But numbers can be very informative if they are treated in the right way. My weight happens to correlate very well with how much fat I'm carrying around my stomach, and hence with what clothes fit me. Certain medical risks go up with weight and with age. You might as well say "Medical risk is just a number". Or even, when you're running a dangerous fever, "temperature is just a number".

Fear causes people to override logic, and that is a good thing in emergency situations. But fabricating fear in order to get people to override logic is not a good thing. Fear also gets in the way in interpersonal situations even if it's not being used as a deliberate means of manipulation. If someone feels under attack it can cause them to override logic, or it causes them to be unable to use logic. Or it causes them to cling on much too strongly to a position they don't really hold. What we need, for productive arguments, is to find ways to build bridges between positions so that people can move, rather than cling on to a position because we've backed them into a corner. And, incidentally, for many people, even if they are open to changing their mind, it might be something they would rather do in private when nobody is looking, like changing clothes.

We are now going to discuss some important ways in which we *should* make use of emotions, beginning with the language we use.

PERSUASIVE EMOTIONS

In Chapter 10 we discussed the starting points of language and the fact that some basic words have to come from somewhere before a language can develop in even a vaguely logical way. Those words can have strong emotional effects on people. When mathematicians choose words for new mathematical concepts they often think very hard about what sort of emotional response they want to elicit. There is something in a name after all, even if "that which we call a rose by any name other would smell as sweet" according to Shakespeare's Juliet. She was, of course, being touchingly naive. Would you be able to sniff a rose seriously if it was suddenly renamed "diarrhoea"? It might take some mental effort.

I had cause to think about the logic and emotions of words when boarding planes recently because American Airlines changed its boarding procedure. Instead of having different "priority groups" boarding before Group 1, it renamed those priority groups with the numbers 1–4, shifting the first normal boarding group to become Group 5. This caused all sorts of confusion because people were not able to listen to instructions, read their boarding pass, or understand the logic of it (or all three). Personally I think, logically, it makes no structural difference as the boarding groups still board in the same order, and surely it doesn't matter whether the groups are called Platinum and Gold or red and blue or banana and frog, or 1 and 2. But apparently some people are very upset about not being in a group called "priority" any more. To them, that word really matters. It is causing an emotional effect and not a logical effect. The people who were upset about not being in a priority group were upset about the word, not about the actual boarding process.

A much more serious version of this mistake over language is a study[1] of young male university students finding that they seem to think forced intercourse might not be rape. The study found that many more men admitted that they had coerced somebody to intercourse than admitted they had raped anybody. One might call it a "false inequivalence" situation where some men think that non-consensual intercourse is not the same thing as rape. It tragically demonstrates how far we still have to go when it comes to education around consent.

The emotional connotations of language can be exploited deliberately, as in the use of the nickname "Obamacare" for the Affordable Care Act (ACA) in the US, as we discussed in Chapter 3. If someone supports ACA but not Obamacare it is painfully clear that they are not evaluating things by their

---

[1] "Denying Rape but Endorsing Forceful Intercourse: Exploring Differences Among Responders". Edwards Sarah R., Bradshaw Kathryn A., and Hinsz Verlin B. *Violence and Gender*. December 2014, Vol. 1, No. 4: 188–193

merits, but by their names. This shows the power of language, again in a "false inequivalence" situation. This is all an example of how something emotional can be used to guide or even manipulate people's thought processes in a way that has nothing to do with logic.

The idea of manipulation has negative associations. If someone is called manipulative that is not a good thing. The world is trying to manipulate us all the time, especially corporations, politicians and the media. One reason to be able to think more clearly is to be able to withstand the onslaught of manipulation that is flooding our world.

However, as with other emotional tools, there are reasons to embrace the power of emotional manipulation if it can help us overcome things in our own lives. Emotional manipulation helped me overcome my fear of flying, when statistics alone wouldn't do it: the logic was no match for my emotions. Emotional engagement was also crucial to me losing weight, as statistics about health risks were not enough – it was not until I became viscerally afraid of becoming morbidly obese that I was able to do it. (It is not acceptable to use this as a technique on other people as it consists of "fat-shaming".)

Sometimes people use gray areas to manipulate us, but we can also use gray areas to manipulate ourselves. I use emotional tricks on myself to motivate myself, overcome mind blocks, stop procrastinating. Perhaps if I were a more logical person I would never lack motivation, get mind blocks, or procrastinate. My mother, the most logical person I know, never procrastinates. But that is a level of logic that is beyond me. Berating myself does not enable me to become logical, and berating others will not make them become more logical either. Thus we need to deal with others and ourselves in an emotional way when we are dealing with the emotional part of ourselves.

In the "post-truth" world, feelings are often presented as facts. Some people accuse others of conflating feelings with facts, especially if they seem to be convinced by something other than logic and evidence. And indeed many people are convinced by personal experience, someone charismatic telling them something, peer pressure, tribe mentality, fear or love. Advertising and marketing drive much of our experience of the world. Marketing is not about demonstrating that a product is better; it is about making people *feel* that a product is better, even if it's essentially the same as all the other ones (or worse).

But these methods of convincing people of things aren't all bad. We generally seem to accept that we become wiser through personal experience, and that there is something valid to be learned from it. Charismatic teachers are an important part of what can make a very good education. Peer pressure and tribe mentality can contribute to change for good as well as bad, for example in the civil rights movement. If people give up smoking because of peer pressure rather than because the evidence says smoking is bad for you, they have at least still given up smoking. But making decisions based on fear is often said to be a negative way of approaching life, and elections based on fear are unpleasant and divisive.

Some fields have become particularly good at invoking emotions to convince people of their message. This might include religious leaders, public speakers, some teachers, advertisers and artists. Science sometimes suffers from the belief that *only* evidence and logic should be used to convey a conclusion. But this is unrealistic. If we are to do that, then we'll have to start by convincing everyone to be persuaded only by evidence and logic. And how do we do that using only evidence and logic, if people are not already convinced by evidence and logic? We end up in something like Russell's paradox.

Moreover, it is unrealistic to think that people can be persuaded by logic alone. In fact, it is in too many cases hypocritical, as scientists themselves are not immune from veering into drawing conclusions based on personal experience or emotions on controversial topics such as women in science. Instead of denigrating emotions in a quest for more rigorous discourse, we should acknowledge their truth and seek to find *the sense in which* there is logic to them.

THE LOGIC IN EMOTIONS

Feelings are not facts. Or are they? It depends what we mean. If I "feel" that $1 + 1 = 3$ that doesn't make it true in normal life. (There is a mathematical world in which it is true, but that's a different matter.) Similarly if I "feel" that someone arrested for a crime is guilty that doesn't make it true.

But there is an important sense in which feelings *are* facts: feelings are always true. If you feel something then the fact that you feel it cannot be argued down by logic. There is rarely any point trying to persuade someone that they "should not" feel something, if they simply are feeling it. Emotions are very powerful at overriding logic.

A more productive process is to find the explanation behind the emotions, uncover the difference between that logic and one you are trying to convey, and use emotions to help bridge that gap. In this way instead of pitting logic against emotions, we separate out emotions and logic in the situation, and only pit logic against logic and emotions against emotions. In a way this is exactly what the process of teaching math well consists of. If a student gives a wrong answer, it rarely helps just to explain the right answer to them. First you have to uncover why they gave the wrong answer, understand the thought process behind it, and somehow convince them that your thought process is more sound.

The reason the emotions are so strong can often be traced back to some fundamental fear. But fear works in strange ways. Sometimes fear works to persuade people of something, and sometimes it doesn't. People who believe the evidence about climate change are typically very afraid about the future of the planet, so feel that it is extremely urgent that we do something about it. Those who do not believe the evidence are typically not afraid of climate change and so don't do anything about it. But why are we not able to make those people afraid enough to want to do something about it? Why by contrast are some people so easily whipped up into a frenzy of fear about refugees even though there is no evidence that refugees are more dangerous than, say, white Americans with guns? Why are some people not at all afraid of gun violence, or at least, not afraid enough to want to have tighter restrictions on guns?

Saying that political views are driven by fear is thus not really an explanation as it doesn't work in all cases – it is an analogy but at the wrong level of abstraction. Perhaps a better analogy between those situations is to do with action. In the case of climate change, taking action against climate change requires some personal sacrifices (better use of resources, which might be expensive). Likewise taking action against guns requires some people's personal sacrifices (giving up their guns). Whereas with refugees, it is being *unafraid* of refugees that requires personal sacrifices – giving up resources to look after refugees, accepting them in the community. Perhaps it is not so much fear that drives these arguments, but personal beliefs around sacrifices.

In this imagined argument we have used a key technique for engaging other people's emotions: the use of analogies.

ANALOGIES FOR ENGAGING EMOTIONS

In Chapter 13 we talked about analogies as the result of performing an abstraction, explicitly or otherwise, and pivoting from one manifestation of it to another. At the simplest level, an analogy is a situation that has something in common with the one you're really discussing. But I think that analogies are hiding something very powerful: a way to try and get people to feel things differently about a situation. Once they feel things differently, they might be able to see the logic differently. The power of the analogy is in doing this via emotions, without having to appeal to anyone's understanding of the logic involved. Unless you are talking to someone already proficient in abstraction and logic, this might be the best you can do.

One way of doing this is to pivot from a situation that they don't seem to feel anything about, to one that they feel a strong emotional reaction to. For example, trying to persuade white women to care more about racism, one might make an analogy with sexism and try to wake up their emotions that way. In a way, this entire technique is a pivot at a more abstract level, from the world of logic to the world of emotions:

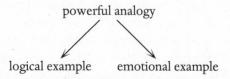

One interesting thing about finding analogies is that when I'm thinking about a good one, I have to think abstractly to find the deep logical argument, and then apply it creatively to another situation to engage someone's emotions. When I am doing this I do feel like I'm using my brain in a highly mathematical manner; it feels like I'm using the same part of my brain. But in arguments in real life the point of it is to enable others to *feel* my point without having to *think* in that highly mathematical manner. In a world where everyone has varying levels of proficiency in abstraction and logic, it is important to be able to find ways of explaining things that bypass that. This is why I make heavy use of analogies when explaining math to students or non-mathematicians, and much less when explaining it in research seminars.

Drawing together the subjects of the last three chapters, we will now show how to use analogies to engage someone emotionally around the issue of prejudice across power relationships. We have talked about the fact that men occupy a position of power over women in society in general. Some people argue that this means men making rude jokes about women is worse than women making rude jokes about men. Or to push it to a greater extreme, that men sexually harassing women is worse than women sexually harassing men. Others argue that this is the same.

There is no right or wrong answer to this, but there is a sense in which the two are the same and a sense in which they are different. The sense in which they are the same is that both consist of "people mistreating other people". At this level of abstraction we have an equivalence.

However, this has forgotten very many details about the situation. We could instead retain the information about power in society, and only abstract as far as "people with power mistreating people without power". At this level, men mistreating women is equivalent to white people mistreating black people, but not equivalent to women mistreating men, or black people mistreating white people:

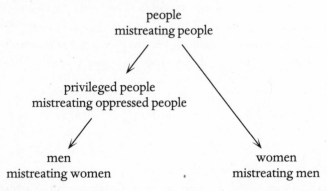

The question has become one of whether or not the intermediate level of abstraction is relevant: that is, whether or not power differential is a relevant factor. I believe it is relevant, but many people insist that it is not. We could try to push them to think about power differentials in general by invoking a more clear-cut example, about which they definitely feel something. For example, a teacher flirting with a pupil. There is a definite problem with power differences here, which is why even if an interaction appears to be consensual, it is illegal between a teacher and a pupil in some countries, just as between an adult and a minor. In the case of minors it is also to do with the absolute inability of a child to consent, but in the case of teachers and pupils the pupil might be over the age of consent but still considered to be unable to consent within that particular power relationship.

We have the following analogy (in fact, an analogous analogy):

a person asking
a person for sex

a person in power asking
someone in their power for sex

a teacher asking   a pupil asking
a pupil for sex   a teacher for sex

I hope everyone would agree that a teacher making advances on a pupil is not equivalent to a pupil making advances on a teacher, because of the power differentials. Similarly a boss trying to seduce an employee is a different situation from an employee trying to seduce a boss, because of the power that the boss has over the employee.

If we can agree that power differentials at least *sometimes* make a difference in some clear-cut cases, then the argument becomes one about gray areas and where to draw the line, if anywhere – which power differentials do make a difference and which ones don't? We can try to persuade people that because white people are so dominant in all levels of power, politics, management, entertainment, and all positions of influence, the position white people collectively hold over black people is analogous to the position a boss holds over an employee.

If they still disagree, it might come down to a question of transferring from the idea of an individual having direct power (boss over employee) and the idea of group power (white people over black people) and whether group power transfers at all to individuals. This is the question of structural racism.

In the process of writing this book and thinking through these arguments very carefully, stripping away layer upon layer to find further abstractions and logical points of view, I have realized how many of these arguments come down to tensions between the idea of individuals and the idea of groups. This applies to the idea of individual vs group responsibility, the extent to which everyone should look after themselves or whether there should be group care. It applies to whether people think a group's treatment by society has any bearing on an individual's. This may go back to a difference in basic personal axioms, in which case we will need to think about how to persuade someone to change their axioms.

We have discussed using analogies to uncover our own personal axioms. But we can also use analogies to uncover other people's personal axioms, to understand why they are thinking the way they are. If we are disagreeing with them because of a different use of logic that is one thing, but if we are disagreeing because of different fundamental principles, it is hard to change those without invoking emotions.

For example, why are some people sometimes so unconvinced by scientific evidence? It might be because they are very invested in not believing it, in which case piling up more evidence won't help: changing their sense of investment will help. Getting to the bottom of why they don't believe it will help, rather than berating them for not believing it.

Once we have uncovered someone's axioms that are at the root of a disagreement, we can start thinking about how to change them. If they are deeply rooted it can be hard, but it could be by experience, meeting people, education, empathy, but in all cases by engaging them emotionally. Our analysis does not tell us how to do that, but at least if we have reached an understanding of really why someone feels something, we are in a better position than if we just think they are stupid.

For many people, emotions and intuition are more convincing than logic. As I have discussed, this is true of mathematics as well, which is why I don't think we should simply scorn the idea. In fact, my whole field of research, category theory, can be thought of as a field that makes precise our mathematical intuitions so that we can do calculations almost entirely by using our intuition, knowing that it will match rigorous logic.

Use of intuition has achieved a lot for mathematicians over the course of history, as long as it is backed up by rigorous justification. So I believe that intuition and emotions can achieve a lot in normal life too, if backed up by logic. Unfortunately while almost everyone feels feelings, not everyone can follow complicated logic. I believe it is therefore incumbent on more logical people to invoke emotional means to make sure logical thoughts are conveyed. This is the subject of the closing chapter of this book.

16

INTELLIGENCE AND RATIONALITY

HOW TO USE LOGIC IN AN ILLOGICAL WORLD

WE HAVE DISCUSSED THE power and the limitations of logic, and the power and limitations of emotions. I am going to conclude with a discussion of how to blend logic and emotions to be a helpfully, persuasively, powerfully rational person. Not just a person who follows the rules of logic, but one who can use logic to illuminate the world of emotional humans.

I will begin by summarizing what I think logical behavior includes and doesn't include, at the most basic level. More subtly, I'll talk about what it means to be not just logical, but *reasonable*. Then I'll go further and describe what I think it means to be *powerfully* logical, when you're not just following the basic rules of logic but also using advanced techniques to build complex logical arguments and investigations, and you are thus able to follow complex logical arguments.

I will show that even if everyone were logical in this way, there would still be plenty of scope for logical disagreement. But most importantly, I will describe what form I think these disagreements would take, and what a logical argument would look like. I wish all arguments took this form. It doesn't mean no emotions would be used. In fact, I'm going to show that even better than being a logical person, I would like everyone to be an *intelligently* logical person. I think this involves not just being logical, but using logic in a way that seeks to help other people, and that this involves a crucial blend of logical techniques and emotions instead of a fight between them. This is what I think intelligence consists of, and it is summed up in this diagram:

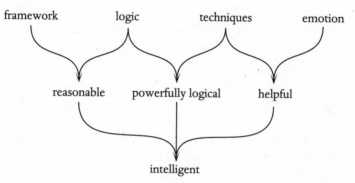

I believe logic is at the core of human intelligence, but that it does not work in isolation.

WHAT IS A LOGICAL HUMAN?

A logical human is one who uses logic. But how? We have seen all sorts of human situations in which logic has limits. To call ourselves logical we should still use logic as far as we can, and no further. Some people see the limitations of logic and conclude that they don't need to use it at all. But this would be like throwing away a bicycle because it can't fly.

I believe that a logical human uses logic, but necessarily has core beliefs that they don't try to justify. This is the starting point of their logic. Then, everything they believe should be attainable from their core beliefs, using logic. Moreover, they should believe everything that follows logically from their core beliefs, and their beliefs should not cause any contradictions.

The idea of core beliefs is analogous to the role of axioms in mathematics, as we discussed in Chapter 11. Believing every-thing that follows logically from your core beliefs corresponds to the logical notion of "deductive closure", which we discussed in Chapter 12. The idea that your beliefs should not cause contradictions corresponds to the logical notion of "consistency", which we discussed in Chapter 9.

If these are the basic principles of being logical, what does it mean to be *illogical*? "You're being illogical!" is used to try and shut down arguments, often by people who like to think of themselves as rational, against people who lead with their emotions (or simply anyone who disagrees with them). But two people can be logical and still disagree, if their logical systems are taking them to different places. Someone who is leading with their emotions might not be able to *articulate* what is logical about their thinking, but that doesn't mean it is actively illogical.

Being illogical means doing things that go against logic, or cause logical contradictions. But I think it is important that these only really count as logical contradictions if they are contradictions *within* your own system of beliefs. This is a crucial point because one person's logic might look like idiocy to another person. I think this is where the battle cry "You're just not being logical" comes from.

Given my definition of a logical person above, there are several valid ways I could judge you to be illogical:

1. Your beliefs cause contradictions, or

2. there are things you believe that you cannot deduce from your fundamental beliefs, or

3. there are logical implications of things you believe that you do not believe.

An example of the first case is all those people who support the Affordable Care Act but not Obamacare. As we've seen, this causes a contradiction because ACA and Obamacare are the same thing, thus those people support and don't support the same thing – a contradiction. An example of the second case might be things that people "just feel", such as when they "just feel" that a relationship is not going to work, or they "just feel" that evolution isn't right, or they "just feel" that it was definitely a vaccination that caused their child to develop autism. An example of the third case is when some men say they don't think health insurance should include maternity cover because they don't think anyone should have to pay for treatment for other people, and they regard maternity cover as only for women (despite the fact that it helps everyone who is born). And yet, they still think prostate cancer treatment should be covered, although that is only for men. In fact, isn't the whole principle of insurance that you pay even when you're not sick, so that everyone can benefit? I think the statement "I don't think anyone should have to pay for treatment for other people" logically implies "I don't believe in insurance". Thus if the man in question still believes in insurance at all, he is being illogical in the third sense. (Of course, we could perform this analogy pivot and discover that probably the principle he believes deep down is that *men* should not have to pay for things that only affect *women*, but it's perfectly fine for women to have to pay for things that only affect men.)

There are a few things to note here. First of all, contradicting someone else's logic doesn't mean you're illogical. Someone might say "It's just not logical, mathematically, to pay $50 to eat something in a restaurant when you could make it at home and spend only $5 on the ingredients." That might be true in *their* system of beliefs, but in my system of beliefs it might well make sense logically to pay for the luxury of having food cooked for me instead of doing it myself. And not having to do the grocery shopping or clean up afterwards. All this doesn't necessarily mean that I am being illogical, it just means that we have different axioms.

The next thing to note is that the question of fundamental beliefs is a gray area. Suppose someone believes, without being able to justify it, that the moon landings didn't really happen. But perhaps they simply think of this as a fundamental belief? It might not seem very fundamental to someone else, but that's a separate question. It comes down to the ability to follow long chains of deductions. We have already mentioned the example of someone saying "I don't believe in gay marriage because I believe that marriage should be between a man and a woman." They may think of "marriage is between a man and a woman" as a fundamental belief, whereas someone else thinks of it as a constructed belief that needs justifying. Likewise if someone believes that you should only vote for someone you truly believe in. One person might think that is an axiom, whereas someone else thinks it needs justification. (I'm amazed that people who think this way ever get to vote at all, but that's a different question.)

The question of whether or not a belief is fundamental enough to count as an axiom is very different from the question of an axiom actually being unreasonable. None of these questions is very clear cut, as we'll discuss shortly. Even the issue of believing something "just because you feel it" could be justifiable if one of your fundamental beliefs is "everything I feel to be true is true". (Incidentally this sounds similar but is very different from saying that feelings are always true.)

Finally note that even the third point, about believing all the things implied by your axioms, gets us into trouble with gray areas. As we discussed in Chapter 12, following the logic inexorably can push us through gray areas to undesirable extremes. For example, if we move in tiny increments, we can logically deduce that it is acceptable to eat any amount of cake at all. The ability to understand gray areas in a nuanced way is an aspect of powerful logic that we will come back to.

The main lesson here is that we need to understand the difference between "illogical" and "unreasonable".

WHAT IS A REASONABLE HUMAN?

I will judge you to be unreasonable if I think your fundamental beliefs are not reasonable. But this might not mean you're contradicting logic, it just means we have some fundamental disagreements. If two mathematical systems have different axioms they do not disagree – they are just different systems, and the best we can do is discuss which system is a better model of the situation in question.

We should acknowledge that what counts as a "reasonable" fundamental belief is a gray area, and is an unavoidably sociological concept: different cultures count different things as reasonable. However, I think a key component of "reasonableness" is that there should be some sort of framework for verification and adjustment.

If one of your core beliefs is that the moon is made of cheese, I would say that this is not reasonable, although it makes for fun fiction (as in Wallace and Gromit's *A Grand Day Out*). But what is my framework for thinking this? First of all, by a logical argument: cheese is a product of milk, and milk comes from animals. How could all that milk product have got into orbit? Secondly, an argument by evidence: people have been to the moon and brought back dust, and it was not cheese.

Of course, there are some people who believe that the moon landings were fake, and that all the evidence about them is part of a huge conspiracy. I would also say this is not reasonable, because I believe in scientific evidence as one of my core beliefs. I will come back to questions of reasonable doubt and skepticism later.

Before we go further we should note that there are some axioms that don't really need to be reasonable: those that are more like personal taste. We are allowed to like and dislike food, like and dislike music. But even those tastes can sometimes be justified further. I used to think my dislike of toast was simply an axiom of mine, but people challenged it so often that I have now explained it more fundamentally by the fact that I don't like crunchy things, and that is because it feels violent to chew them. You might think I'm absurd, or ridiculously sensitive, but I think it's within my rights as a reasonable person to decide I don't like the feeling of chewing something crunchy.

Aside from outright contradictions it is hard to talk about what counts as reasonable core beliefs without being stuck floating in a space of relativism: you might worry that I can only call someone's beliefs unreasonable relative to mine, at which point they can call mine unreasonable relative to theirs, and indeed many arguments take this futile form in which both sides call the other unreasonable and no progress is made.

Setting aside questions of personal taste, there is one criterion for reasonableness that I think has a chance of not being relative, and the clue is right there in the word "reasonable": are your beliefs open to being reasoned with? That is to say, are you open to changing them? Do you have a framework for knowing when it is time to change them? Are there any circumstances at all under which you would change them?

In one of my favorite moments of *Macbeth*, Macduff is trying to persuade Malcolm to come back from exile and fight Macbeth for the throne of Scotland. Malcolm has a clever and wise way of discerning whether or not this is a trap to lure him to danger. He starts portraying himself as a terrible person, and describes what a cruel and evil king he would be. He needs to see whether Macduff's support of him is rational or not. If it is rational, then in the face of Malcolm's admissions he will withdraw his support. If he does not withdraw his support, Malcolm will conclude that the support is not rational and he is therefore not to be trusted. In the event, Macduff despairs and cries, "Oh Scotland, Scotland!" and withdraws his support, determining to leave Scotland himself forever. Because Macduff withdraws his support in the face of the supposed new evidence showing how unsuitable Malcolm is to be king, Malcolm is reassured that the support is rational.

I think this openness to changing one's conclusions or axioms in the face of evidence is an important sign of rationality. If someone continues to support a person or idea or doctrine regardless of further and further evidence then this is a sign that the support is blind rather than rational. There is a difference between loyalty and blind support, and a difference between healthy skepticism and science denial. I think it's an example of fuzzy logic. Loyalty means not changing your support over minor issues. Blind support means not changing your support over major issues, or any issues at all. Of course, a question remains over what counts as "major" and "minor" issues.

Here are some things I have changed my mind about over the years. I have already mentioned compulsory voting in Chapter 13. I also now support liberal arts education because I see that this can happen either informally (as in the education I received) or formally (as in the US system). I now support a more active form of feminism because I see that the passive form was not achieving the change I want to see. I (grudgingly) support getting up early, because I've discovered it helps me lose weight, possibly for hormonal reasons. And I believe in doing things for myself, not just for other people, because I see that if I neglect myself I reduce my ability to do things for other people.

If I examine these cases carefully I see that I have changed my mind about axioms from a combination of logic, evidence and emotions. Even if it's not explicit, there is some kind of framework there.

FRAMEWORKS

We have discussed the framework that math and science have for deciding what to accept as truth. For math it's logical proof. For experimental science the framework consists of finding evidence. It is based in statistics, which means that scientists are required to find evidence to back up a theory to a good level of certainty. The framework then says that if new evidence arises to overturn that level of certainty or even point in a different direction, science changes the theory accordingly. This is very different from the kind of "theory" where you just make something up because you feel like it.

We can examine something similar for the framework of news reporting. Reporters are supposed to gather information to back up their story, according to a certain framework of accountability. It is less rigorously defined than in science, but there are still standards to do with cross-checking and reliability of sources. Again, this is very different from the kind of "news" where someone just makes something up. In both cases the report might turn out to be wrong, but in the first case there is a procedure for discovering it is wrong and retracting it, whereas in the second case there isn't.

This is the crucial difference between erroneous reporting and "fake news". Unfortunately the term "fake news" has been appropriated by some people to mean, more or less, "anything I disagree with". If a newspaper retracts an article because they find that their sources turned out to be unreliable or misinformed, some people are likely to shout "Fake news!" However, at least the newspaper has a framework and procedure for verification of its reports. It is always unfortunate when something only turns out to be wrong after publication, but this happens in science despite much more rigorous validation processes, so is bound to happen in journalism, which works with less rigor and much more time pressure. It is important for the rational among us to maintain the distinction between statements arrived at via a framework and those without. It is tempting to try and distinguish between "facts" and falsehoods, but if you follow logic carefully you should find it difficult to say for sure what a fact is. The best we can do is have a statement verified according to a well-described framework, and an allowance for the fact that the framework might later find it to be wrong.

At this point we are once again in danger of getting caught in a loop, because there are reasonable and unreasonable frameworks. If "reasonable" is defined according to having a "reasonable framework", have we actually got anywhere or are we just making a cyclic definition?

I think this is why people can disagree so much about what counts as reasonable and what doesn't: because the notion of what counts as a reasonable framework is sociological, just like the notion of what counts as a valid mathematical proof turned out to be sociological. One group of people thinks that the scientific method is the most reasonable framework, whereas another group thinks it is a conspiracy. One group thinks that the Bible is the most reasonable framework, and another group thinks it is a piece of fiction.

This is why one of the few things I can come back to as a sign of unreasonableness is if someone is absolutely unprepared to change their mind about something. This often takes the form of hero-worship, and I believe it is very dangerous to rational society.

THE MYTH OF HEROES, SUPERSTARS AND GENIUSES

Skepticism is an important part of rationality, and loyalty is an important part of humanity, but both become dangerous when taken to extremes. Blind skepticism and blind loyalty arise when there are no conditions under which someone will change their mind – or that the conditions are so extreme that they might as well not exist.

For example, a climate change denier might say they'll believe in global warming if the average temperature on earth rises by $10°C$ in one year. That hardly counts as being "open" to changing one's mind because it's a bit like saying "OK I'll believe in that if hell freezes over". Deniers of evolution will probably not change their minds no matter what quantity of evidence is produced supporting it, so scientists should probably stop using evidence as a way of trying to persuade them, and try using emotions.

Blind loyalty can be dangerous in another way. When people support a person regardless of anything at all, it can lead to that person gaining a sort of cult status as a superstar or "genius". Unconditional support sounds like a noble thing, but really should be in some kind of gray area like so many other things. How badly does someone have to behave for you to stop supporting them? Parents are often thought to show unconditional love for their children, but this might be pushed close to or beyond its limits if the child grows up to be a mass murderer.

That is an extreme case, but we see less extreme cases around us all the time in the form of people who exploit their power. When someone starts feeling like they have unconditional support of people who revere them as some kind of "genius", they might start behaving badly, knowing that they can count on the blind loyalty of their followers. This can happen in all fields, including science and academia, music, TV and film, and the restaurant industry. It contributes to a culture in which exploitation and harassment are widespread, and so I think we should stop it. Of course this is not a simple issue. At what point should we withdraw our support for someone? It comes back to the difference between "minor" issues and "major" issues and is yet another gray area.

GRAY AREAS OF REASONABLENESS

Gray areas have been popping up repeatedly throughout this book. They seem to be everywhere, and I think we need to accept that and deal with it, and acknowledge that being rational involves accepting that some things are rather fuzzy. For example, many things are "just theories" but that doesn't make them all equally trustworthy, or equally dubious – it depends what sort of framework has been used to establish that theory. Similarly if a large group of people or sources agree with each other, that doesn't necessarily mean that there is a conspiracy, but it might – it depends, again, what sort of framework has been used to establish that agreement.

There are many degrees of trust and skepticism that we can show towards theories, sources, experts and evidence. It's not just about trusting something or not, there's a huge gray area in between.

Should we believe scientific "experts" or not? At one extreme, some people think that scientists are all in a conspiracy with each other. At another extreme, some people regard science as absolute and unassailable truth. Against science, some people think that trusting science means you're an unthinking sheep, and that intelligent people are always skeptical about everything. They cite scientific theories from the past that have turned out to be wrong. In favor of science, some people think that those who are skeptical of science are being irrational and using emotions instead of logic. Both sets of people are liable to think the others are being stupid, and this is not a helpful situation.

I think we should acknowledge that there are gray areas everywhere. For skepticism there is healthy skepticism and blind skepticism and everything in between. For trust there is also healthy trust, blind trust, and everything in between. I would say that healthy skepticism and trust come from, again, a well-defined framework, including evidence and logic.

Blind trust and blind skepticism might actually look on the surface quite similar to the healthy versions. The two versions might be equally fervent. But I call someone's trust or skepticism blind if they can't justify it to many steps. I can't justify my belief in science to the end (because there is no end) but I can keep going for a while: I believe in the system of the scientific framework because it has checks and balances; it is self-reflective and self-critical; it is a process rather than an end result; it has a framework for updating itself and has known occasions when it has found itself to be wrong and corrected itself.

Some people think that admitting you're wrong is a sign of

weakness, or that changing your mind is a sign of indecision. But I think both of these are an important sign of having some framework for your trust and skepticism. That, to me, is a sign of a more powerful form of rationality.

POWERFUL RATIONALITY

Being rational is a start, but is not enough. You can avoid illogic but still not get anywhere, like someone who travels safely by simply never going anywhere. That is different from travelling safely while going all over the world. Being *powerfully* rational means not just using logic and avoiding logical inconsistencies, but using logic to build complex arguments and gain new insights.

Throughout this book I have discussed logical techniques and processes that I think contribute to powerful rationality. This starts with abstraction, which is what enables us to use better logic in the first place. I think it then has three main components: paths made of long chains of logic, packages made of a collection of concepts structured into a new compound unit, and pivots using levels of abstraction to build bridges to previously disconnected places.

Abstraction is the discipline of separating out relevant details from irrelevant ones, and finding the principles that are really behind a situation in such a way that we can try to apply logic.

It is then important to be able to follow a long chain of deductions, both forwards and backwards, and not just a single step like a child who can't get further than "If I don't get ice cream I will scream." We follow logic forwards, to comprehend all the consequences of one's thinking, and backwards, to construct and understand complex justifications of things. This includes being able to axiomatize a system down to very fundamental beliefs, rather than just believing things because you do, and it also includes being able to understand *someone else's* beliefs. If you can't follow long chains of logic backwards you will be stuck taking almost everything you believe as a fundamental belief. This isn't exactly illogical, but it's not very insightful either, and hardly leaves open the possibility for fruitful discussion. "Why do you think that?" "Because I do." I think powerful rationality involves being able to unpack your reasoning down to a very small number of core beliefs, and being able to answer "Why do you think that?" down to very deep levels. Just like mathematicians should be able to fill in their proofs to as deep a level of "fractalization" as anyone might ask, we should be able to do that with our beliefs too.

Building inter-related ideas into compound units is an important source of power in logic. The ability to think of a group of things as one unit is something we do naturally every day, when we think of a family, a team, or compound nouns for animals: a flock of birds, a swarm of bees, a herd of cows. We think of a school (and all the people making it up), a business, a theater company. I much prefer using singular verbs with these compound nouns, as I really am thinking of them as single units. I will say "My family is going out for dinner" rather than "my family are going out for dinner".

Packaging complex systems into single units should not mean forgetting that the system is made of individuals. Powerful rationality involves understanding the way in which the individuals are interrelated, forming the whole system, as we saw in Chapter 5. After looking at those huge diagrams of interconnected causations you might despair that the situation is so complicated. However, if we develop our logical power so that we are able to comprehend and reason with those complex systems as single units, then it will no longer seem complicated. Gray areas are encompassed in this idea about complex systems, as they consist of situations where instead of getting a simplistic yes or no answer out, we have a whole range of related answers on a sliding scale. This is like having a range of probabilities for different possible outcomes, rather than trying to predict one outcome. It might seem hard to understand a range of probabilities rather than one prediction, but a powerfully rational person will then develop the skill of understanding the more difficult concept, rather than giving up and resorting to the simplistic one. The same is true of gray areas.

We tend to look for a single cause or a single answer to a question. One way to find one cause for a complex situation is simply to ignore all the others, as people frequently do when blaming an individual for a complicated situation. However, another way to find a single cause is to package the whole system up and be able to regard that as "one cause".

This enables us to think more clearly and also move to different levels of abstraction. We discussed at length in Chapter 13 how analogies consist of using abstraction to make pivots to other situations. I believe powerful rationality involves great facility at moving between different levels of abstraction to make different sorts of pivots, to move between different contexts and see many points of view.

Powerful rationality involves being able to separate axioms from implications, which is related to being able to separate logic from emotions. This doesn't mean suppressing one or the other, but understanding what role each is playing in a situation, and what each is contributing. It involves finding logical justifications or causes of emotional facts, including other people's. This leads me to an even more important aspect of rationality: how to use it in human interactions.

I think there is something even better than being a powerfully rational person, and that is being an *intelligently* rational person, which is someone who is not just powerfully rational, but uses that power to help the world, somewhat in the way that the best superheroes use their superpower to help the world. And the best way I think that we can use this superpower to help the world is to bridge divides, foster a more nuanced and less divisive dialogue, and work towards a community that operates as one connected whole.

INTELLIGENT RATIONALITY

Life doesn't have to be a zero-sum game, where the only way to win is to ensure that someone else loses. People who think it does are usually trying to manipulate other people whom they think they can beat. I may sound rather optimistic, but there are abundant examples of situations where people collaborate for the greater good, rather than compete. This is the essence of teamwork and communities, and perhaps the very essence of humanity. We are not, after all, each living in a cave by ourselves, but living in communities at many different scales: families, neighborhoods, schools, companies, cities, countries, and even, with any luck, cooperation between countries.

I believe in a slightly modified version of Carlo M. Cipolla's theory of intelligence in *The Basic Laws of Human Stupidity*. He defines stupidity and intelligence according to benefits and losses to yourself and others.

If you benefit yourself but harm others, you are a bandit. If you benefit others but hurt yourself (or incur losses), he calls this "unfortunate", though I might rather say you are being a martyr. Both of these make life into a zero-sum game. On the other hand there are people who hurt others and themselves at the same time, as in the prisoner's dilemma. Cipolla defines this as stupidity. The remaining possibility is to help yourself and others at the same time, and Cipolla defines intelligence to be the quadrant of mutual benefit:

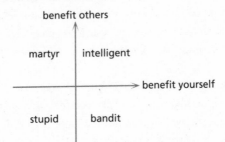

This is an eye-opening definition of intelligence, involving nothing to do with knowledge, achievements, grades, qualifications, degrees, prizes, talent or ability. I like it, and it is this form of intelligence that I will use to describe intelligent rationality. Intelligent rationality is where you don't just use logic, and you don't just use it powerfully, but you use it in human interactions to help everyone. The aim should be to help achieve better mutual understanding, to help others and yourself at the same time. If you are only using logic to defeat someone else's argument and promote your own, that is the intellectual version of being a bandit.

Intelligent rationality is about using logic in human interactions, and so it must involve emotions to back up logical arguments in all the ways I have already described. Without this, I don't believe we have any serious chance of reaching mutual understanding with those who seem to disagree with us. Conversely, intelligent rationality should involve being able to find the logic in someone else's emotional response as well as our own, rather than just calling emotions wrong.

For example, when I was offered a chance to move to Chicago I was perplexed because rationally it was obviously the best choice for me, but emotionally I felt reluctant. In order to understand this dissonance I wrote a list of weighted pros and cons, and I discovered why I was confused: in favor of the move were a small number of really enormous points, but putting me off the move was a huge list of minor details. I had emotionally become swamped by the huge quantity of minor details. Once I had discovered the source of my fear I was able to reduce it, and in the end I made the decision with no hesitation, and no regrets.

Another example is when I eat far too much ice cream although I know it's going to make me feel ill later. I could tell myself I'm just being illogical, but it's more nuanced than that: I am prioritizing short-term pleasure (delicious ice cream) over medium-term pain. That's not illogical; it's a choice, and once I see it as that, I am sometimes able to make a different choice.

Arguing and reasoning with oneself is a good first step, but what about arguing with others? What should we do about people who disagree with us?

WHY LOGICAL PEOPLE STILL DISAGREE

It is important to acknowledge that logical people can still disagree. It doesn't mean that one person is being illogical, although that might be the case. Possibly both people are being illogical. It also doesn't mean that both people are being stupid. Logical people might disagree because they are starting with different axioms.

For example, perhaps one person believes in helping other people, and another person believes that everyone should help themselves. Those are different fundamental beliefs, but neither is illogical. In fact, I would say it's a false dichotomy: I believe that everyone should help themselves, but that some people are privileged with more resources to help themselves than others, so we should all also try to help those less privileged than us.

Logical people can also disagree because of the limits of logic. Once we've reached those limits there are many different ways we can proceed, depending on what means we choose to help us once logic has run out. Often it is a case of picking a different way of dealing with a gray area, or picking a different place to draw an arbitrary line in a gray area. If one person accuses the other of not being logical, it may be the case that *neither* person is being entirely logical because the scope of logic has run out.

I think an important aspect of being more than just basically logical involves being able to find the sources of these disagreements, and this involves using logic more powerfully, to have better arguments.

GOOD ARGUMENTS

What I want to see in the world is more *good arguments*. What
do I mean by that? I think that a good argument has a logical
component and an emotional component and they work
together. This is just like the fact that a well-written mathematical paper has a fully watertight logical proof, but it also has
good exposition, in which the ideas are sketched out so that we
humans can feel our way through the ideas as well as understanding the logic step by step. A good paper also deals with
apparent paradoxes, where the logical situation appears to
contradict our intuition.

The important first step in a disagreement is to find the true
root of the disagreement. This should be something very close
to a fundamental principle. We should do this by following long
chains of logic in both our argument and theirs. We should try
and express it in as general a principle as possible, so that we
can fully investigate it using analogies.

Next we should build some sort of bridge between our
different positions. We should use our best powers of abstraction and pivots to try and find a sense in which we are really just
at different parts of a gray area on the same principle.

We should then engage our emotions to make sure we
engage theirs and understand them where they are, and then try
and edge slowly to where we can meet. This will include
finding out what, if anything, would persuade them to change
their mind. We also have to show that we are reasonable
ourselves, and that we are open to moving our position too, as
we should be if we are reasonable. If we really understand their
point of view we may discover things we didn't know that
really do cause us to move our position or even change our
mind.

I think a good argument, at root, is one in which everyone's
main aim is to understand everyone else. How often is that
actually the case? Unfortunately most arguments set out with

the aim of defeating everyone else – most individuals are trying to show that they are right and everyone else is wrong. I don't think this is productive as a main aim. I used to be guilty of this as much as anyone, but I have come to realize that discussions really don't have to be competitions. If everyone sets out to understand everyone else, we can all find out how our belief systems differ. This doesn't mean that one person is right and the other wrong – perhaps everyone is causing a contradiction relative to everyone else's belief system; this is different from people causing a contradiction relative to their own belief system. Unfortunately too many arguments turn into a cycle of attack and defense. In a good argument nobody feels attacked. People don't feel threatened by a different opinion, and don't need to take things as criticizm when they're just a different point of view. This is everyone's responsibility, and if everyone is an intelligently powerful rational human being, everyone will assume that responsibility for themselves. In order to achieve that, we all need to feel safe. Until everyone is in fact that intelligent, those who are should try to take responsibility for helping everyone to feel unattacked. I try to remind myself as much as possible in any potentially divisive situation: it's not a competition. Because it almost never is, in fact, a competition.

A good argument does invoke emotions, but not to intimidate or belittle. A good argument invokes emotions to make connections with people, to create a path for logic to enter people's hearts not just their minds. This takes longer than throwing barbed comments at each other and trying to throw the "killer shot" that will end the discussion, and I think this is right. Logic is slow, as we saw when we looked at how it fails in emergencies. When we are not in an emergency we should have slow arguments. Unfortunately the world is tending to drive things faster and faster, with shorter and shorter attention spans meaning that we are under pressure to convince people in 280 characters, or in a pithy comment that can fit in a few words around an amusing picture, or a clever one-liner – correct or otherwise – so that someone can declare "mind = blown" or "mic drop". But this leaves little room for nuance or investigation or finding the sense in which we agree along with the sense in which we disagree. It leaves no time for building bridges.

I would like us all to build bridges to people who disagree with us. But what about people who don't want bridges? People who really want to disagree? This is a meta problem. First we have to persuade people to want those bridges, just like motivating people to want to learn some mathematics before we have any hope at all of sharing it.

As humans in a community, our connections with each other are really all we have. If we were all hermits living in isolation humanity would not have reached the place it has. Human connections are usually thought of as being emotional, and mathematics is usually thought of as being removed from emotions and thus removed from humanity. But I firmly believe that mathematics and logic, used in powerful conjunction with emotions, can help us build better and more compassionate connections between humans. But we must do it in a nuanced way. We have seen that black and white logic causes division and extreme viewpoints. False dichotomies are dangerous in the divisions they cause, both in the mind and between people. Logic and emotions is one of those false dichotomies. We should not pit ourselves in futile battles against other humans with whom we are trying to coexist on this earth. And we should not pit logic against emotions in a futile battle that logic can't win. It's not a battle. It's not a competition. It's a collaborative art. With logic and emotions working together we will achieve better thinking, and thus the greatest possible understanding of the world and of each other.

ACKNOWLEDGMENTS

FIRST AND FOREMOST I would like to thank Andrew Franklin and all at Profile Books for their extraordinarily far-reaching support. Heartfelt thanks are also due to Lara Heimert, TJ Kelleher and all at Basic Books. It is a great honor to have publishers on both sides of the Atlantic who believe in me and push me to keep developing as an author. For this book I must particularly thank Nick Sheerin for his brilliant editorial help.

I owe many thanks to my students at the School of the Art Institute of Chicago. Their intellectual energy and social conscience have pushed me to use abstract mathematics for more and more social issues, which has directly led to this book. I would also like to thank everyone at the School. It is a revelation to work for an institution that so utterly values all aspects of my work.

None of this would happen without the support and inspiration of my parents, my sister and my little nephews Liam and Jack, who are not as little as they were when I thanked them in my previous book.

Thanks are also due to my wonderful friends whose insights, anecdotes and arguments push me to think more clearly and galvanize me to put my thinking to good use. Special thanks are due to those whose thoughts I have specifically referred to in this book: my PhD supervisor Martin Hyland, my English teacher Marise Larkin, my math teacher Andrew Muddle, and also Will Boast, Oliver Camacho, Daniel Finkel, Jessica Kerr, Sally Randall and Barbara Polster. And I am always grateful to Sarah Gabriel for being my ongoing beacon during brain fog.

Chapter 13 is dedicated to Gregory Peebles, with whom I have the best and most compassionate arguments.

INDEX

Eugenia Cheng is the Scientist in Residence at the School of the Art Institute of Chicago and an Honorary Fellow of the University of Sheffield. The author of three books, she lives in Chicago, Illinois.